Charles Dixon

Lost and vanishing birds

Being a record of some remarkable extinct species and a plea for some threatened forms

Charles Dixon

Lost and vanishing birds
Being a record of some remarkable extinct species and a plea for some threatened forms

ISBN/EAN: 9783337112837

Printed in Europe, USA, Canada, Australia, Japan

Cover: Foto ©berggeist007 / pixelio.de

More available books at **www.hansebooks.com**

FRONTISPIECE
PLATE 1

GREAT BUSTARDS

LOST

AND

VANISHING BIRDS

Being a Record of some Remarkable
Extinct Species and a Plea for
some Threatened Forms

By CHARLES DIXON
AUTHOR OF
"THE MIGRATION OF BIRDS" "CURIOSITIES OF BIRD LIFE"
"THE NESTS AND EGGS OF BRITISH BIRDS"
ETC. ETC. ETC.

WITH TEN PLATES BY CHARLES WHYMPER

LONDON: JOHN MACQUEEN
MDCCCXCVIII

PREFACE

ONE of the saddest features of civilisation is the disappearance of so many beautiful and curious creatures from this world of ours. From all parts of the earth the same story comes; and we now seem to be within measurable distance of a time when wrecks and remnants of once compact and indigenous assemblages of organisms will be all that remain to us, and such a thing as a complete fauna will be unknown. This is not only a crime, but the violation of a sacred trust which we hold for posterity. Civilisation has already ground away under its merciless heel most of the faunal facies of Europe; Asia fares but little better, and is fast being reduced to the same state; Africa is being rapidly depleted of all its most curious and striking forms of animal life; Australasia is a wretched object lesson of civilised man's exterminating progress; whilst North America has already lost some of its ancient types, and is fast losing the remainder: South America alone retains its prehistoric fauna in greatest completeness, although even here the sad work of extermination has commenced. Birds have suffered severely in this general spoliation, and their extermination and persecution furnish material for some of the saddest chapters in the annals of ornithology.

In the present volume an effort has been made not only to focus in a popular form our knowledge of the species we have lost and are still likely to lose, but to excite a greater interest in the protection of birds, particularly in those species, at home and abroad, that are more or less threatened with extermination at the present time.

So far as British birds are concerned, we have dealt with all the recently extinct and threatened species; but of course it would be utterly impossible, within the limits of this small volume, to treat exotic species with the same fulness. We have, however, carefully selected a few of the most interesting and desperate cases on which to hang our plea for the better protection of all. Some of the most interesting extinct species have also been included.

PREFACE

We are convinced that much of the effort now being made on behalf of doomed or threatened birds is misdirected; and if the present work not only helps in some measure to devise more rational methods, but also excites a wider sympathy for those vanishing species, its principal purpose will have been attained.

C. D.

PAIGNTON, *October* 1897.

CONTENTS

LOST AND VANISHING BIRDS

	PAGE
INTRODUCTION—THE EXTERMINATION OF SPECIES	13

PART I

LOST AND VANISHING BRITISH BIRDS

Lost British Birds

SAVI'S WARBLER (*Locustella luscinioides*)	43
THE SPOONBILL (*Platalea leucorodia*)	48
THE BITTERN (*Botaurus stellaris*)	54
THE CRANE (*Grus cinerea*)	60
THE GREAT BUSTARD (*Otis tarda*)	67
THE AVOCET (*Recurvirostra avocetta*)	73
THE BLACK-TAILED GODWIT (*Limosa melanura*)	78
THE BLACK TERN (*Sterna nigra*)	83
THE GREAT AUK (*Alca impennis*)	87

Vanishing British Birds

THE BEARDED TITMOUSE (*Panurus biarmicus*)	98
THE ST. KILDA WREN (*Troglodytes hirtensis*)	104

CONTENTS

Vanishing British Birds—continued

	PAGE
The Hoopoe (*Upupa epops*)	108
The Osprey (*Pandion haliaetus*)	113
The Kite (*Milvus regalis*)	119
The Common Buzzard (*Buteo vulgaris*)	125
The Golden Eagle (*Aquila chrysaetus*)	130
The White-tailed Eagle (*Haliaetus albicilla*)	136
The Honey Buzzard (*Pernis apivorus*)	142
The Marsh Harrier (*Circus æruginosus*)	147
Montagu's Harrier (*Circus cineraceus*)	152
The Hen Harrier (*Circus cyaneus*)	157
The Dotterel (*Eudromias morinellus*)	162
The Kentish Sand Plover (*Ægialophilus cantianus*)	167
The Ruff (*Machetes pugnax*)	173
The Red-necked Phalarope (*Phalaropus hyperboreus*)	179
The Roseate Tern (*Sterna dougalli*)	185
The Great Skua (*Stercorarius catarrhactes*)	190
Some Threatened British Species	196

PART II

LOST AND VANISHING EXOTIC BIRDS

Lost Exotic Birds

The Mamo (*Drepanis pacifica*)	211
The Dodo (*Didus ineptus*)	215
The Solitaire (*Pezophaps solitaria*)	220
The Pied Duck (*Camptolaimus labradorius*)	226
Pallas's Cormorant (*Phalacrocorax perspicillatus*)	231
Some Other Extinct Forms	234

CONTENTS

Vanishing Exotic Birds

	PAGE
The Carolina Paroquet (*Conurus carolinensis*)	237
The Owl Parrot (*Strigops habroptilus*)	242
The Passenger Pigeon (*Ectopistes migratorius*)	245
The California Vulture (*Pseudogryphus californianus*)	250
The Heath Hen (*Tympanuchus cupido*)	254
The American Turkey (*Meleagris americana*)	256
The Aldabran Rail (*Dryolimnas aldabranus*)	261
The Kiwis (Apterygidæ)	266
Struthious Birds: Ostriches, Rheas, Emus, and Cassowaries	271
Some Threatened Exotic Species	283

LIST OF ILLUSTRATIONS

PLATE		
I. Great Bustards		*Frontispiece*
II. Fenland in the Olden Days.	*To face page*	48
III. Avocets	,,	73
IV. Great Auks	,,	87
V. Bearded Tits	,,	98
VI. The Kite	,,	119
VII. The Golden Eagle	,,	130
VIII. The Mamo	,,	211
IX. The American Turkey	,,	256
X. Kiwis	,,	266

INTRODUCTION

THE EXTERMINATION OF SPECIES

PERHAPS few readers are aware (unless they be experienced and professed zoologists) how very sensitive species are to any changes in their surroundings: on the one hand, quick to take advantage of anything in their favour; on the other hand, as readily injured by adverse conditions. These latter may be of the most varied character, and make their influence felt in a very complicated or indirect manner, the relations not only between one species and another, but with their environment, being most complex. Many instances might be given to illustrate how complex are the relations, not only of one species to another, but to the environment of those species, or, in other cases, to the utter dependence for existence of species upon their neighbours. During the lapse of unnumbered ages, all living things have been (and still continue to be) unceasingly striving, under the influence of certain well-recognised laws, to adapt themselves to more or less constantly changing conditions of existence. What is popularly known as the "balance of nature" is the primal result of these incessant efforts of organisms, one acting upon the other in countless ways, to maintain a place in the ranks of struggling life. We can very forcibly illustrate these remarks by quoting one or two classical instances recorded by Darwin. Certainly one of the most complex of these is that which illustrates the intricate connection between, and interdependence upon, such widely different organisms as a carnivorous animal and a scented yet lowly flower. Perhaps every reader may be aware that certain flowers absolutely depend upon the visits of insects to fertilise them. They cannot produce seed without such visits; and in a great many instances this fertilisation can only be accomplished by a certain species of insect. Now, one of our commonest flowers, the red clover, is largely, perhaps we might almost say entirely, fertilised by our little friend the humble-bee. If these bees do not visit the clover flowers, those flowers are sterile and produce no seeds. But the humble-bees have a deadly enemy in the field-mice, which destroy, it has been computed, no less than two-thirds of their nests and combs. The mice in their turn are destroyed by cats, Owls, Kestrels; so that in localities where the enemies of mice are common the bees have more chance of multiplying, and the flowers a correspondingly greater facility for fertilisation. The abundance of clover in a district may therefore depend upon the number of cats, of Owls and Kestrels! Take another instance. Darwin has recorded some very curious effects produced by the planting of several hundred acres of Scotch fir on a large heath in Staffordshire. In a quarter of a century the change produced in the vegetation was very remarkable, plants having appeared or disappeared in obedience to the altered conditions, whilst many other organisms were undoubtedly similarly affected. One more instance must suffice, and this we may quote from Darwin's great work on *The Origin of Species*: "In several parts of the world insects determine the existence of cattle. Perhaps Paraguay offers the most curious instance of this, for here neither cattle nor horses nor dogs have ever run wild, though they swarm southward and northward in a feral state; and Azara and Rengger have shown that this is caused by the greater numbers, in Paraguay, of a certain fly which lays its eggs in the navels of these animals when first born. The increase of these flies, numerous as they are, must be habitually checked by some means, probably by other parasitic insects. Hence, if certain insectivorous birds were to decrease in Paraguay, the parasitic insects would probably increase; and this would lessen the number of the navel-frequenting flies. Then cattle and horses would become feral, and this would greatly alter (as, indeed, I have observed in parts of South America), the vegetation; this, again, would largely affect the insects; and this, as we have just seen in Staffordshire, the insectivorous birds; and so onward, in ever-increasing circles of complexity. Not that under nature the relations will ever be as simple as this. Battle within battle must be continually recurring with varying success; and yet in the long-run the forces are so nicely balanced that the face of nature remains for a long time uniform, though assuredly the merest trifle would give the victory to one organic being over another."

Most, if not all, organisms are therefore so delicately adapted to their environment, that they quickly become sensitive to the least disturbing

element, either for good or for evil, profiting
readily by the former, and being adversely affected
by the latter, even to the extent of more or less
rapid extinction. Numerous instances might be
given to illustrate how readily certain species have
profited by the decrease, say, of their natural
enemies, or the initiation of easier conditions of
existence; and, on the other hand, how disastrous
have been the effects of similarly changed conditions
acting in a directly opposite manner. We have,
for instance, much cause to regret the rapid
increase of the House Sparrow, partly due to the
wholesale slaughter of birds of prey, and partly
to the exceptional facilities for shelter, abnormal
reproduction, and the constant and abundant
supply of food, due to the march of modern
civilisation and the spread of agriculture. We
have equally to regret the disappearance from our
avifauna of such species as the Great Bustard and
the Crane. Bird lovers may well deplore the final
disappearance of such magnificent species from our
islands, due indirectly, to some extent, to the changed
conditions of the century now drawing to a close,
but more to the growth of sport, the increase of
gunners, armed with more deadly weapons, and
the rapid multiplication of the avaricious class of

collector. We may safely conclude that senseless persecution and wanton slaughter must be held primarily responsible for the loss of the Great Bustard, aided by alteration in the methods of tillage. Drainage and enclosure of waste lands, and the changed conditions due to increase of population, and possibly the spread of railways and other industries that have broken the seclusion and almost primeval peace of many a favoured haunt, must also be held responsible for the bird's disappearance, as well as indiscriminate shooting and egg-stealing.

So far as we are at present able to ascertain, the disappearance of species from the world may be more or less directly traced to the agency of man, and primarily of civilised man. We cannot recall to mind a solitary instance in which the extermination of a species within historic time has been exclusively due to any extra human agency. Species and individuals, of course, are constantly striving one against the other in the battle of life; incessantly struggling to maintain a place in the ranks of existing forms—here gaining an advantage, there losing ground, as the conditions of existence may vary to their disadvantage or in their favour. The extermination of species under such conditions we know must have taken place, as the records of palæontology unquestionably demonstrate, and there can be little or no doubt is in actual progress around us now; but the process is so gradual, and the difficulties of direct observation and calculation so immense, that we entirely fail to perceive it. Some slight indication of the exterminating force of unfavourable natural conditions may be derived from the effects, say, of a severe winter, or an abnormal season of drought or wet, or unusual lowness of temperature, upon birds, for instance; but these adverse circumstances are never sufficiently prolonged for us to remark the absolute decimation of a species, and apply but to a circumscribed area. On the other hand, the extermination due to man's interference with the balance of nature is immeasurably more rapid, and its results in the majority of cases are only too sadly apparent. Many, indeed, are the instances which might be quoted in support of these statements. Uncivilised man, so long as he uses primitive weapons, apparently makes little or no evil impression upon continental faunæ, the slight tax upon them being amply met by the normal increase of the species concerned, but in islands the case has been different, as will be seen in future pages.

In all parts of the world island species have been the greatest sufferers and the most easily exterminated, owing partly to the comparatively limited number of individuals composing them, and in a great measure to their very specialised and localised conditions of existence rendering them acutely sensitive to any adverse influence. The extinction of a great many intensely interesting forms—in the present volume we shall confine ourselves to birds alone—may be said to date from that period when the early explorers were scouring the seas in quest of undiscovered countries, and when remote uninhabited islands were either permanently colonised or periodically visited for the supplies of fresh food and water that they may have chanced to furnish. In most cases the visits of civilised man to these islands has had sooner or later a disastrous effect upon the avifauna which is or was usually peculiar to them. Man not only destroyed many of these wonderful bird-types for food or other purposes, but brought about their gradual extirpation less directly in other ways—by burning off the undergrowth or clearing the forests, and by introducing various domestic or predatory animals to which the peculiar, and in many cases flightless birds, or their still more helpless eggs and young, fell easy victims. In these remote times, the small amount of interest taken in what we may call *living* science, when zoologists attached no importance whatever to the geographical distribution of species, nor to the equally significant phenomena of island faunæ and floræ as bearing upon the question of the evolution of specific forms, may reasonably be urged as an excuse for the want of some efforts being made to preserve for posterity these interesting and valuable relics of an ancient past. But this extenuating circumstance cannot be pleaded as an excuse for the almost universal work of extermination that has been going on steadily and surely through the present century; even after the publication of the discoveries of Darwin and Wallace, that not only changed the entire process of zoological research, but brought out in vivid suggestiveness the importance of those forms which civilised man has been (consciously or not makes no difference) doing his best to stamp out. More especially do we allude to the senseless crime of extirpation which has been committed in New Zealand and other antipodean lands, where species after species has passed away, and others are still surely following, without any rational efforts being made to save them. Zoologically this region is perhaps the most interesting in the world. It contains many species and types unknown elsewhere, many of them archaic forms, relics of a once perhaps dominant fauna replaced by more highly specialised forms, and only preserved to us at all by that isolation which has eventually wrought their doom. Not only have these species been directly destroyed by man, but the senseless practice of "acclimatisation" has here been pursued in all its crass stupidity. Man by his silly meddling methods, and his tampering with that balance which nature so delicately established and kept true, has worked sad havoc amongst indigenous species. By way of illustration: first rabbits were transported to the Antipodes, and then, when they became a pest,—as was long foreseen by naturalists,—ferrets, stoats, and weasels were introduced as a futile attempt to exterminate them. But these predatory creatures, instead of materially lessening the rodent plague, attacked the helpless fauna, especially the flightless birds, with results that can only possibly end in the complete extinction of these interesting forms. This introduction of exotic species, where successful, almost invariably ends sooner or later in disaster to some members of the indigenous fauna with which they are brought most closely in contact; and we may here take the opportunity of protesting most strongly against that introduction of various foreign birds into our islands which has been suggested by more than one naturalist, philosophic enough, one would think, to realise the inevitable consequences, more especially so with such unhappy examples of "acclimatisation" before them. The House Sparrow, to quote but a single instance, was imported into America as a welcome novelty and souvenir of the Old Country; it has now become such a pest that a fruitless war of extermination is almost everywhere waged against it, and the bird in not a few places has succeeded in ousting indigenous and far more interesting and useful species.

In the absence of all historical evidence, and with nothing but tradition and legend to guide us, it is impossible to form any correct estimate of the number of avine species that has been exterminated by uncivilised races of mankind. We have, however, some comparatively recent evidence furnished by the Maoris of New Zealand, whose traditions relating to certain species of gigantic wingless birds of that country, known as "Moas," are of exceptional interest. From information which has been gathered from the Maoris, there seems little or no reason to doubt that their ancestors were well acquainted with these huge birds in a living state, and that at one time the Moas frequented both islands in abundance. The Maoris hunted them for food, and as the birds must have been comparatively helpless, and possibly of low fecundity, the improvident natives eventually exterminated them, shortly before the arrival of civilised man in New Zealand. Possibly another instance of avine extermination by savage man is presented by the Mamo (*Drepanis pacifica*), of the Sandwich Islands, that is said to have been killed for its yellow plumage, which was used to embellish the state robes of chiefs. We are also informed by Dr. Forbes, that since the Chatham Islands were colonised by Maoris and Europeans some fifty years ago, the birds have lamentably decreased in number, and the constant persecution of every sort of bird and living thing by the natives is producing the certain extermination of all the indigenous species. But the natives in this case may only be following the white man's example, or tempted by the price which is often offered for a rare bird by collectors. The Moas undoubtedly owed their extinction to the Maoris,

INTRODUCTION

who found in them an easily procurable supply of food, but for the subsequent decimation of the New Zealand fauna Englishmen themselves are solely to blame. There can be little doubt that one of the most deadly exterminators of the indigenous birds of New Zealand is the rat. The brown rat was introduced into the islands during the very earliest days of their settlement, and, as usual wherever it finds its way, it took readily to its new home and multiplied apace. Then came the introduction of stoats and weasels, and between them these bloodthirsty little animals have worked sad havoc amongst the indigenous birds, most of which are, or were, not only exceptionally tame and unsuspecting in a land where there were few or no enemies, but made their nests in places readily accessible to these four-footed invaders. When brought under the influence of such changed conditions, most birds seem powerless to avert their threatened extinction, and instances are excessively rare in which a species has altered its habits to escape from an entirely new danger. One such instance we may, however, quote—that of the Samoan Pigeon (*Didunculus strigirostris*). This species, in order to escape the cats which threatened speedily to exterminate it, is said to have taken to nesting and roosting in high trees, with the beneficial result that its numbers are now steadily on the increase again. It is gratifying to have Mr. W. W. Smith's assurance that in certain parts of New Zealand some of the rat-threatened birds—Honey-eaters and others—are again increasing in number, as conditions are proving less satisfactory for their four-footed foe, and the clearing away of the lower bush is depriving the rat of a favourite haunt. Another fruitful cause of extinction is the importation by settlers, from sentimental motives, of certain birds from Europe, notably the House Sparrow, which have succeeded in crowding out many indigenous species. Dogs, cats, goats, and hogs, when introduced into small islands, have also exterminated many helpless avine species, especially ground birds and those in which the power of flight was limited or even absent.

Comparatively few people are aware how rapidly and upon what an enormous scale the spread of civilisation is working changes and making serious gaps in the fauna of the world. Civilisation, wherever it spreads, sooner or later affects the wild creatures of the invaded area, and in most cases the change has been attended with disaster to the fauna. Islands do not suffer alone, for even the great continents are now rapidly being depopulated of their larger or most helpless birds and beasts. The work of extermination may in many cases be a longer one than it has proved to be on many islands, but the final results are just as inevitable. In the Polar regions the seal and the whale (to quote but a couple of instances) have been reduced almost to a state of extinction; in warmer lands the zebra and the giraffe of Africa, in fact all the big game of that continent, is rapidly being exterminated; in America the buffalo and other large animals are threatened with a similar fate. Every year civilised man (and to a great extent savage man follows his example) is becoming more and more utilitarian, and species after species is threatened as its economic value becomes recognised. Millions of birds must be killed annually for decorative purposes; crocodiles, alligators, lizards, and many other wild creatures, formerly despised, have been found to yield valuable products; and if the fashion or craze lasts, the species affected ultimately verges on extinction. Wherever civilised man and his animal satellites penetrate, the fauna suffers, and the longer he remains the more disastrous his influence becomes; so that it requires no very severe strain upon the imagination to picture a time when all the larger wild birds and beasts, all the exceptionally helpless ones of the earth, must perish, or exist only as specimens in our museums, or as phantom records in our scientific literature. This will be a serious outlook for the biologist of the future, and the matter has long been sufficiently important to warrant some strong steps being taken to avert as far as possible such a vast calamity. After all, we only hold the fauna of the world in trust, and it is but our bare duty to posterity to hand that fauna down as intact as we found it, or as nearly so as the reasonable exigencies of life will admit.

We now come to consider the question of extermination in a partial sense, and more especially as it relates to our own islands. The species with which we are therefore concerned now are those that have become extinct in some parts of their range, although they still survive in other areas. Here, again, islands present us with the most significant and important instances of recent extinction, although many continental examples might be cited where birds have been extirpated in some localities although continuing to flourish in others. The Passenger Pigeon of America and the Francolin of Europe may be quoted as cases in point. A very large percentage of the birds whose absence from the British Islands as breeding or indigenous species we have now to deplore, probably could not have been preserved to us had the most elaborate means for their protection been devised. They were victims to the results of advancing civilisation and improvement—destined by the altered conditions of existence that such changes involved, to disappear from certain areas in which it became impossible for them to survive. On the other hand, there are certain lost species that might still have continued to find a place in our avifauna had reasonable protection been granted to them. These, too, have passed from our area never normally to return. There are certain other interesting species still left to us, but extermination awaits them in the by no means distant future, unless steps be speedily taken to preserve them.

The unscientific reader may naturally ask why comparatively so few birds have become extinct in the British Islands, where the influence of civilisation has been so prolonged and so acute, whilst so many have suffered in New Zealand and other remote islands whose colonisation has been relatively so recent. This apparent anomaly admits of a very easy explanation. Islands that have from a variety of causes, which we need not here stay to discuss, remained in a state of great isolation, are generally found to be inhabited by a fauna, or the relics of a fauna once more widely dispersed, or have developed a variety of species by the aid of their long-enduring isolation from all allied forms. It thus happens that these remotely isolated spots have gradually become possessed of a fauna more or less peculiar to themselves, species being found on them that are not found anywhere else. But, on the other hand, islands that are not so isolated, either being situated close to continents, of which it is certain they formed a geologically recent part, or are located in seas in which uninterrupted intercommunication with the nearest land masses is maintained by normal migration across them, or the various fortuitous methods of dispersal, have few or no such opportunities for establishing a peculiar fauna, and consequently preserve their biological homogeneity. The British Islands are a capital example of the latter class of islands, and their avifauna is almost exactly identical with that of the adjacent continent, and is subject to very similar conditions. But two birds are peculiar to them: one of these, the Red Grouse, is carefully preserved from extinction for the sport it yields; and the other, the St. Kilda Wren, had long maintained its place even on a few isolated rocks, until in an evil day its specific difference was detected, and now the greed of collectors threatens soon to extirpate it as effectually as other methods did the Dodo and the Great Auk. In Britain, then, we had no peculiar or flightless birds, no species so tame from its unfamiliarity with man, for civilisation to extirpate, although we had certain others—individuals of widely dispersed continental species—that bred in our islands, many of which have vanished or are gradually going, more perhaps than the average reader is likely to suspect. We cannot too strongly assert, as having a vital bearing upon the whole question of extermination, that the supply of birds, even in such a favourable locality as the British area, situated as it is so closely to continental land, is inexhaustible. If we kill off our native contingent, especially of resident or breeding species, there is no reason whatever to console ourselves with the belief that other individuals will arrive to replace them. If such were really the case, the Great Bustard, the Spoonbill, the Crane, and other vanished species would be dwellers in our land to-day; for there are plenty of these birds across the Channel, almost within view of the white cliffs of England. But individual birds are closely confined to certain areas, and to these they keep with fatal pertinacity, so that, if we destroy all the individuals in one area, the chances are that that area will remain depopulated for ever. The record of extermination in the British Islands abundantly proves the truth of this assertion; for in every case where the native stock has been exhausted, the species has dropped out of our fauna completely, unless introduced by man, as the sedentary Capercaillie was. No bird of strictly migratory habits that has been exterminated in the British Islands will ever return to them again, notwithstanding any and every effort that man may make to reinstate the species. The sedentary Bustard might be induced to take up its quarters with us again, but the migratory Crane under no circumstances whatever will ever return as our summer guest. Bearing these facts in mind, it behoves us to guard jealously what few large birds remain to us, and in the case of vanishing species to see that they are carefully preserved, especially during the breeding season, when their numbers may in time gradually increase.

There can be no doubt, of course, that the great alterations which have been made in many districts, especially in reclaiming waste lands, have literally destroyed the haunts of many of our larger birds. These changes were inevitable; but when we bear in mind how attached individual birds are to their accustomed haunts, we cannot help feeling that if protection had been given at the right time, some at least of these big birds might have been preserved to us even if in a semi-domesticated condition. We have surely the familiar instance before us in so many continental towns and villages, of the White Stork returning year by year to rear its young on the houses and mosques, or the Hoopoe stalking sedately on the dunghills of the Arabs, regarded by the inhabitants of these countries with no more curiosity than we evince for the Swallows and Starlings nesting on our own dwellings. We may rest assured that the birds would stay with us as long as existence were possible, if we left them unmolested. It is too late now to retain many of our lost birds, but there are others left that would appreciate protection, and pass their harmless, nay, even useful lives in our midst. The gunner should be restrained, the bird-catcher warned off, even the collector forbidden. Legislation on behalf of our vanishing birds has been most beneficial, and might, of course, be of greater service; but we would rather see our favourites preserved by sentiment and kindly feeling than protected by Act of Parliament. We should like to see lessons on the uses and economy of birds become part of our national education, and kindness to birds inculcated and fostered in every school in the land.

On the other hand, as a set-off against the many interesting species that we have lost for ever, it is gratifying to know that the spread of cultivation and the improvement of waste land, so disastrous to the larger birds, has favoured the increase and dispersal of considerable numbers of the smaller species. Many of these latter birds are songsters of varying merit, and these have followed the horticulturist and the agriculturist, so that many districts are now made glad with song which formerly were silent. The boom of the Bittern has died away with the disappearance of marsh and fen; the song of the Passere is heard in its place. This, in a measure, is some compensation for our loss. In some districts, however, many of the smaller birds have been ruthlessly depleted by the gunner and the snarer; and we can name localities where such species as Goldfinches, Bullfinches, Hawfinches, Wood Larks, Nuthatches, and Kingfishers are either altogether exterminated or fast becoming so. Certain intelligently framed Amendments to the Acts for the Preservation of Wild Birds, and the establishment of proper machinery for the enforcement of the existing law, should remedy the evil. The wholesale destruction of the nests and eggs of the smaller birds that goes on in most country districts must have a most injurious effect upon the species, and is even worse than the destruction of the birds themselves. Eggs to some extent are now protected, but the law in most places is utterly ignored.

A few words here seem appropriate upon the practice of shooting those odd birds that accidentally visit our islands from time to time. Now, of the four hundred or so of avine species which comprise what is popularly known as the "list of British birds," nearly one half are practically abnormal visitors to our shores, lost and stray individuals, as a rule, far from their proper area of distribution, and doomed sooner or later to "die without issue." Without in any way being understood to countenance or defend the indiscriminate destruction of birds purely for the sake of killing, we maintain that the capture of these wanderers does not injuriously affect the species in the slightest degree, but, on the other hand, is a direct service to the science of ornithology. Their capture is often of great importance, and the thanks of all systematic ornithologists are due to the collector of every abnormal avine visitor to British shores. We often hear of a burst of indignation greeting the publication of such a capture, but wrath of this kind is as untimely as it is out of place. None of these wanderers will ever succeed in establishing the species in our area; avine colonisation does not depend upon such methods, and if every rare abnormal visitor were left severely alone, the net result would be precisely the same. But a certain amount of discrimination is absolutely necessary, especially in spring. For instance, the Hoopoe arrives on our southern shores so frequently in spring, that there is the possibility of these visits being normal. The bird should therefore be left to rear its young in peace if so minded; and I would have every rascal pilloried that dared to shoot one of these curious and charming creatures. But such species as the Bee-eater, the Yellow-browed Willow

Warbler, White's Thrush, and the Desert Wheatear may be shot without compunction; for the capture of a hundred of these birds in England would be less injurious to the species than the death of a single pair at their normal breeding-grounds or winter quarters; in fact, it is even the more merciful course to shoot them, for it prevents their ultimate death from starvation or worse. All these abnormal visitors are already dead to their species, and their capture is not only advisable but perfectly justifiable.

One word in conclusion. There are few subjects concerning which more nonsense has been written, or which are more surrounded with maudlin sentiment, than the "extermination" and "slaughter" of birds. In season and out of season we are being constantly reminded by well-intentioned people, we do not doubt, that this bird or that is threatened with extinction, or being ruthlessly butchered. The capture of a "rare bird" is often the signal for an outburst of misplaced indignation from these well-meaning faddists, whose ill-timed diatribe too often not only defeats its object and brings ridicule upon themselves, but is apt seriously to injure a cause whose welfare every naturalist should have at heart—the protection of our native avifauna, and the actual preservation of threatened species. Let not these remarks be misunderstood; for we yield to no one in our desire to see our feathered friends and favourites shielded from harm, or more heartily condemn their often useless and unnecessary slaughter. But let us put our own house in order first: there is much to do at home in the intelligent protection of our native birds, and in guiding public opinion, before we turn elsewhere. It may be perfectly true that abroad certain birds are sorely persecuted for their plumage; but the facts are often grossly exaggerated; and the inconsistency of these ignorant "humanitarians" repels rather than attracts sympathy, and defeats its own ends. Doubtless there will always be fair women ready to adorn their persons and enhance their charms by the aid of borrowed plumes, all Leagues and Societies notwithstanding, and in moderation and humane discrimination who shall say them nay? but the crusade against the abuse of the practice would be far more effective if more rationally and sensibly conducted. We offer these words of advice out of no ill-feeling to these well-meaning folk, and assure them of our sympathy and support in every movement for the intelligent preservation and protection of the birds. In some respects accredited

INTRODUCTION

collectors and scientific men are as much to blame in decimating a species as the milliner and his fashionable lady patrons. Birds, many of them local and scarce to a high degree, are being indiscriminately collected in the name of science. Naturalists, of all people, should ever seek to protect, never heedlessly to destroy.

We will now proceed to notice in detail not only lost and vanishing British birds, but some of the principal exotic species already extinct or threatened with extermination.

Part I

LOST AND VANISHING
BRITISH BIRDS

LOST BRITISH BIRDS

SAVI'S WARBLER

(*LOCUSTELLA LUSCINIOIDES*)

IN many respects Savi's Warbler is a very interesting little bird. In the first place, it may safely be regarded as the most fleeting known species that has ever occupied a place in the British fauna; for it was not discovered to be a British bird at all until about the year 1819, and in less than forty years it had, so far as can be ascertained, become extinct in our islands, the last specimen having been obtained in 1856. Savi's Warbler becomes still more interesting to English naturalists from the fact that the species may be said to have been first discovered in the British Islands, although its specific distinctness was not declared until four years after this event, when in 1824 Savi gave it a name. All the evidence we possess relating to the British distribution of Savi's Warbler indicates the very restricted nature of its habitat. So far as is known, this Warbler was confined to the fens of Norfolk, Cambridgeshire, and Huntingdonshire. Like the Dartford Warbler, it was therefore one of our most local species—a significant fact, as we shall shortly learn.

In our opening chapter we have pointed out the usual fate that overtakes species localised on islands, when their conditions of life are seriously changed. Precisely the same remarks apply to Savi's Warbler; its very localness (as was equally the case with the Large Copper Butterfly, a denizen of the same fenland wastes) was the principal cause of its rapid final extinction. No direct war was waged against it, but its few chosen haunts were reclaimed and brought into cultivation, so that existence in them became impossible. Had Savi's Warbler been more widely distributed, like its congener the Grasshopper Warbler, for instance, there can be no reasonable doubt that it would have been in existence as a British species to-day. It is a rather remarkable fact that such a species should have had so restricted a distribution in our islands, and one that seems to suggest that its numbers had been steadily diminishing for years before the species was discovered. Its fate should serve as a warning, for we have other excessively local species in our midst—the Marsh Warbler, the Dartford Warbler, the Chough, and the Red-necked Phalarope, to name but a few—which may become extinct as rapidly, not necessarily through the destruction of their favourite haunts, but from direct persecution. Savi's Warbler also sadly confirms the fact previously dwelt upon, that the supply of birds (whether sedentary or migratory species) in a district is by no means inexhaustible, and in the present case must have been a very limited one indeed. This Warbler still breeds in the fens of Holland, but from similar causes—the drainage of its aquatic haunts—is much less common than formerly. All allowance being made for the excessively skulking habits of Savi's Warbler, there can be little likelihood of its ever being detected in our country again, and no human agency can ever restore it to our avifauna. We will now proceed to give a few brief particulars concerning the life history of this vanished species.

Savi's Warbler appears everywhere to be a singularly local bird, and breeds in various suitable districts of Central and Southern Europe, and in

North Africa in the swamps of Algeria and Morocco. It is a summer visitor to the south of France, to Spain, Italy, Austria, and Central and Southern Russia. The birds that breed in the Kirghiz Steppe area and in Turkestan are possibly sub-specifically distinct. The only winter quarters of Savi's Warbler appear to be in Egypt and in the oases of the Sahara. The haunts of this Warbler are apparently confined to reed beds. The bird is said not to be so shy as its congener the Grasshopper Warbler, but is skulking and wary enough if alarmed, taking refuge amongst the reeds. It may often be seen running mouse-like up the stems of the reeds to the feathery crown, then dropping again into the cover to repeat the action on another stem. Sometimes it pauses on the crown of a reed to utter its exceedingly monotonous song, which closely resembles that of the Grasshopper Warbler—more musical, perhaps, but far less powerful. This song is uttered both by day and by night. The call-note is described as a harsh *krr*. Savi's Warbler, like most other reed Warblers, is a somewhat quarrelsome bird, and ever ready to drive away a rival or an intruder from its particular haunt.

The nesting season of this Warbler is in May

and June. We are informed by Professor Newton and others that the nest of Savi's Warbler was well known to the Fen men, although they were unacquainted with the parent birds. The nest is carefully concealed amongst the aquatic vegetation from a few inches to a few feet from the ground, and is a well-made, deep, cup-shaped structure, composed almost entirely of the flat, ribbon-like leaves of *Glyceria*. The eggs—from four to six in number—vary from white to pale buff in ground colour, sprinkled and freckled with light brown and violet grey underlying markings. Both birds are said to assist in incubation, and but one brood appears to be reared in the season. The food of this Warbler consists principally of insects and their larvæ.

Savi's Warbler is a sombrely arrayed little bird, having the general colour of the upper parts uniform russet brown, darker on the quills. The under parts are pale buffish brown, becoming nearly white on the throat and the centre of the belly, and pale chestnut on the under tail coverts. The female closely resembles the male in colour. The total length of the bird is about five and a half inches.

THE SPOONBILL

(*PLATALEA LEUCORODIA*)

ALTHOUGH the Spoonbill is still an abnormal visitor at irregular intervals to our islands, it must now be regarded as another of our lost British birds. We do not share the recently expressed opinion of an eminent naturalist, that these accidentally occurring individuals would doubtless once again take up their residence amongst us; for what we already know of the laws of avine dispersal is diametrically opposed to such a proceeding. These odd wandering Spoonbills that from time to time pay us their uncertain and irregular visits are migrants out of their proper course, not pioneers in quest of pastures new; and these, we doubt not, will gradually cease to be noticed in England at all as the bird becomes extinct in Holland, its last stronghold in North-Western Europe, and where most of its breeding-

PLATE II.

FENLAND IN THE OLDEN DAYS

places are gradually being destroyed. We have ample evidence to show that the Spoonbill was formerly widely if locally distributed over the southern and eastern portions of England and in the south of Wales. In England, in the olden days, the Spoonbill was known by the names of "Popeler," "Shovelard," and "Shoveler," whilst the Duck known to us by the latter term was then called a "Spoonbill." We learn many interesting facts about the Spoonbill from ancient records—that it used to build in company with Herons in Norfolk and Suffolk: that earlier still there were colonies of Spoonbills established at Fulham in Middlesex, and in some of the woods of West Sussex. There are also records of this species breeding in trees in Pembrokeshire. The last breeding-place of the Spoonbill in England of which we appear to have any record was at Trimley in Suffolk. This was about the year 1670. It is difficult to assign any reason for the Spoonbill's extinction in this country. The reclamation of fens and marshes is not a sufficiently satisfactory explanation, for the Spoonbill appears to have been equally at home in high trees; a more feasible cause of its disappearance may have been the destruction of timber and the breaking up of land for building purposes, together with that direct persecution which such a curious and conspicuous bird would be sure to invite, especially as the improvement in and the carrying of firearms became more general. The fact also that the birds were left unprotected during the breeding season, although the taking of the eggs was punished with a severe penalty, could not fail to have a disastrous effect upon the species. Had equally stringent measures been taken for the preservation of the birds during this critical period, the Spoonbill might still have been numbered as an indigenous English species to-day. We understand that in Holland the bird is now strictly preserved in some of its ancient strongholds, which we hope may result in retaining this handsome species in the Dutch *ornis* for many years to come.

In Europe, in addition to Holland, the Spoonbill breeds in Southern Spain, in the valley of the Danube, in the delta of the Volga, and in the Aral basin. Eastwards in Asia we trace it as a breeding species, in Asia Minor, Turkestan, Western Siberia up to 48° north latitude, Southern Dauria, the Amoor Valley, South - eastern Mongolia, and southwards over the whole of India and

Ceylon. The Spoonbill also breeds throughout Africa, south to the Soudan, and the Dahalak Archipelago in the Red Sea. It is a winter visitor to Arabia. The Spoonbill is only a summer visitor to Europe, arriving in April, and leaving in September and October. Its favourite summer haunts are swamps, especially those near the sea, the shallow reed and rush clothed margins of lakes, and the dense thickets of willow and alder trees on the submerged banks of large rivers like the Danube and the Volga. The Spoonbill is a gregarious species, and not only lives in societies, but frequently mingles with other Herons, Ibises, and Cormorants. Its habits are very similar to those of its allies. It has the same sedate walk, and may often be seen standing in the shallows or on the topmost branch of a tree quite motionless. Like most large birds, it is somewhat shy, but at its breeding-places will pass to and fro in silent flight above the head of the intruder. It is not known to utter a note of any kind, but frequently makes a sharp clapping sound with its bill after the manner of a Stork. Its food principally consists of small crustaceans, insects, and molluscs, the bird searching for them in the Duck-like way for which its broad spatulate bill is so admirably adapted. It also captures small fish, frogs, and, it is said, eats various vegetable substances. The Spoonbill probably pairs for life, and yearly returns to the same haunts to breed. The nests in some districts are placed upon the ground, in others upon low bushes, in others again upon lofty trees. Nests made in the branches are larger and more elaborate than those placed upon the ground. When in the latter situation it is often nothing but a low heap of broken reeds; when in trees and bushes, often a large mass of sticks, a foot high and a yard across, the cavity containing the eggs being usually lined with dry grass. The old nests are often repaired year by year, just as is the case with Rooks. The eggs of the Spoonbill are four or five in number, coarse in texture, white in ground colour, sparingly spotted and blotched with reddish brown, and still more sparsely with underlying markings of grey. They are subject to much variation in size. But one brood is reared in the season.

The Spoonbill has the general colour of the plumage white, suffused or stained with yellow on the neck and crest, the latter (a nuptial ornament) formed of a bunch of narrow pointed and drooping plumes. The spatulate bill is

THE SPOONBILL

black on the basal portion, shading into yellow at the tip; the legs and feet are black. The female resembles the male in colour. The total length of this species is about thirty-two inches.

THE BITTERN

(*BOTAURUS STELLARIS*)

THE Bittern is another species that visits us more or less irregularly on migration, but one which is unfortunately lost to our indigenous avifauna for ever. We do not for a moment believe that these odd birds which reach us will ever attempt to settle in the British Islands as permanent residents. The old race of indigenous Bitterns has passed away. These we have every reason to believe were sedentary; whilst those that visit us to-day do so to winter in our islands only, just as is the case with so many other species, some individuals of which, however, are indigenous and breed with us, as, for instance, the Starling, the Snow Bunting, the Song Thrush, and the Goldcrest. Now, we think it may be taken as one of the primary conditions of avine dispersal, that species do not increase their range with a winter movement, or attempt to colonise for breeding purposes areas they may visit on autumn migration. Normal dispersal is the result of range expansion in spring for purposes of reproduction. That being so, we hope the reader will understand that the Bitterns still visiting us are not seeking in any way to extend their breeding area; that they are descendants of those individuals which increased the range of the species across our islands or from a British base, probably when the North Sea was an extensive marshy plain, and are in the habit of returning here to winter or to pass over our area to more southern districts. Introduction by man might succeed in reinstating the Bittern as a British bird, as it did the Capercaillie; but we need not foster any hopes that the species will ever settle here without such aid, however carefully we may preserve these visitors, or whatever inducements we may offer them to do so. Be all this as it may, the Bittern should not be shot at all in this country, or the few that still continue to visit us in winter or on passage may ultimately be exterminated, and the bird cease to be a "British" one in any sense of the term. The Bittern, from all accounts, was pretty generally and commonly distributed over the British Islands "in the days of long ago,"—that is to say, in suitable localities.

These were the swamps and bogs and fenlands, and the drainage of these was one of the principal causes of the bird's extermination in our land. Possibly the esteem in which it formerly used to be held as a table delicacy may also have been responsible for its decrease, together with the improvements in and increase of firearms. As might naturally be expected, the Bittern lingered long in the Fen districts—the last eggs being taken in Norfolk in 1868. It is also said that a young bird was caught in the Broad district so recently as 1886, but whether it was *bred* there is not absolutely certain. The bird also continued to breed in Ireland down to the early part of the nineteenth century, but now it is only known as a winter visitor, as it is elsewhere. The Bittern has a wide distribution outside the British Islands, being found in all suitable localities throughout Europe, Asia, and Africa. It does not penetrate very far north, being unknown in Norway, and only visiting Sweden up to the 60th parallel. In Russia it is found up to latitude 62°; in Asia apparently not beyond latitude 57°.[1] To Europe

---

[1] Seebohm obtained a skin in the valley of the Yenisei in latitude 64°, but the evidence is not conclusive that the bird was obtained there.

the Bittern is principally known as a summer visitor, though some few birds winter on the northern shores of the Mediterranean.

The habits of a bird of such a secretive nature as the Bittern are very difficult to observe or understand, and little surprise can be felt at the amount of mystery and superstition that has surrounded them. The bird's haunts are also most difficult of access, being by preference the vast reed beds and swamps. Although apparently migrating in companies, at other times the Bittern is a remarkably solitary bird, and one that delights to skulk amongst the cover, taking wing with reluctance, and depending largely for safety upon the resemblance of its brown pencilled plumage to the vegetation in which it is hiding. The Bittern is apparently more nocturnal in its habits than its allies the Herons, and during the pairing season its singular awe-inspiring cry or "boom," peculiar to the male, is heard at intervals all through the night — a weird, indescribable double call said to be produced as the bird inhales and exhales its breath and stands with neck outstretched and bill pointing upwards to the sky. So curious is the sound, that the country-folk used to say the bird produced it by blowing into a reed or burying its long spear-shaped bill in the mud—

Like as a Bittern that bumbleth in the mire.

The Bittern is seldom seen upon the wing, and flies in a slow, deliberate manner, seldom for any great distance at a time, and always apparently anxious to hide itself as quickly as possible. Less rarely still is it observed to alight in a tree. Like all the Heron tribe, the Bittern has a voracious appetite, feeding chiefly on fish, frogs, and aquatic insects, and occasionally on small animals; eels a foot or more in length have been taken from its stomach. Upon the ground the Bittern is able to run through the dense reeds with marvellous celerity, its long slender feet enabling it to cross the marshy ground with ease. Of the pairing habits of the Bittern but little is known. The bird is a somewhat early breeder, the eggs being laid in April and May—sometimes towards the end of March. The nest is made upon the ground in the reeds and other aquatic vegetation, and is little more than a heap of rotting reeds, flags, and other herbage. The four or five eggs are brownish olive or buff. The female is said to incubate these for the most part, and but one brood is reared in the season. The Bittern is just as solitary during the breeding season, each pair keeping to a particular haunt. The young are said to remain in the nest until they are able to fly.

The Bittern has the general colour of the plumage buff, irregularly vermiculated and pencilled on the upper parts and streaked on the lower parts with black, which is the uniform colour of the head and nape; the feathers of the neck are elongated into a very conspicuous ruff. Bill and bare space before the eye greenish yellow; legs and feet light green; irides yellow. The female and young do not differ to any great extent in colour from the male; and the total length of an adult bird is about twenty-eight inches, sometimes a trifle more or less.

THE CRANE
(*GRUS CINEREA*)

THERE can be little doubt that formerly the Crane was one of those species which not only bred in the British Islands, but visited them in considerable numbers to pass the winter. Whether the individuals that bred in Britain were residents does not, however, seem very clear. Possibly these birds came in spring to breed in the British marshes, and retired south again in autumn, their places being taken during the winter by migratory individuals from still more northern haunts, as the Woodcock is thought by many naturalists to do to-day. Whatever were the real facts, there is ample evidence to show that the Crane formerly bred commonly in the British Islands. Its principal strongholds appear to have been the fens and marshes of East Anglia and the bogs and morasses of Ireland. There can be little doubt that the Crane began to diminish as a breeding species in the British area towards the close of the twelfth century, continuing to do so through the three following centuries, and finally ceasing by the end of the sixteenth century. Simultaneously the extermination of the Cranes that visited these islands exclusively for the winter appears to have been in progress. As might naturally be expected, the indigenous or breeding birds were the first to go; and there is evidence to show that the Crane still continued to visit the fens for the winter long after it had ceased to breed within our limits. During the latter half of the seventeenth century the Crane was only known to Willughby and Ray as a winter visitor in large flocks to the Lincolnshire and Cambridgeshire fens; but these must have become exterminated early in the eighteenth century, for in 1768 Pennant informs us that the bird was quite unknown in those counties. From that time to the present the Crane can only be regarded as an irregular and abnormal visitor on migration to various parts of the British Islands, sometimes occurring in exceptional numbers, as in the year 1869, and drawn here, we may rest assured, by no nostalgic impulse, but driven to our island shores by the exigencies of their annual journeys to destinations far remote from them. What was the cause of this noble bird's extinction in our islands? Probably a potent cause was the drainage of its marsh and fenland haunts. We know that the eggs and nestlings of the Crane were protected by law; but perhaps these steps may have been taken when the bird was already fast vanishing from the land: however, the fact that the parent birds were not included rendered any such provision futile in the extreme. In any case, we well know that legal protection of such a character was unable to save the bird from extinction; and we should feel disposed to attribute its disappearance as a breeding species to the destruction of its nesting haunts and to the killing of the old birds during the breeding season, whilst undue persecution may have also assisted in reducing the numbers of the birds that came into our area for the winter only. A bird so large and conspicuous, such a noble prize, would be sure to be unduly harassed by the fowler; and as the favourite haunts became smaller and more accessible to man, in spite of its wariness the poor Crane would dwindle in numbers, winter after winter, until all were gone. The worst of it is, in this case, too, the Crane is absolutely lost to us, it can never be reinstated into our fauna; the odd birds that visit us are abnormal migrants, and we may safely rest assured that the old stock of indigenous individuals and regular winter migrants has long passed away. We might add, in concluding this historical survey of the Crane as a British species, that remains of the bird have been found in the "kitchen middens" of Ballycotton in County Cork.

The Crane has a very extensive range, being a breeding species in all suitable localities throughout Europe and Northern Asia, and wintering in various parts of Southern Asia and Europe, and in Africa as far south as the northern limits of the intertropical realm. In Europe it visits the Arctic regions to breed, as well as many localities in South Russia, Turkey, the Danube area, Austro-Hungary, Italy, Andalusia, Germany, Poland, and the Baltic Provinces. In Asia it does not go quite so far north (the Arctic Circle in the extreme west, latitude 60° farther east), but in the south it breeds in Turkestan, the Baikal area, and the valley of the Amoor. Its winter home in Asia is in Persia, Palestine, South China, and Northern India. Three years ago Dr. Sharpe separated the Asiatic individuals as *Grus lilfordi*, on the ground of their presumed paler coloration, but their specific distinctness has not been very generally recognised by naturalists.

The migrations of the Crane are by no means the least interesting portion of the bird's life history. These migrations extend from the tropics to the Arctic regions, and are performed at vast heights and by great numbers of individuals flying in company. Cranes begin to cross the Mediterranean into Europe as early as February and March, often passing over certain spots in successive flocks, the birds trumpeting to each other as they go. The Crane appears to migrate by day alone, and the flocks on passage either assume the form of a V or a W, or each bird flies in single file. The haunts most favoured by the Crane are extensive swamps, full of lakes and quaking bogs, mingled with higher and drier ground clothed with coarse herbage, heath, and scattered bushes. Although many of these places are entirely surrounded with forests, the Crane shows no partiality for trees. Few birds are more wary or more quick to detect advancing enemies, and the stalking of a Crane in its open haunt is almost an impossibility. Except on passage, the Crane spends most of its time upon the ground, walking with graceful steps, and wading into the shallow water in quest of food. The flight is strong and well-sustained, the big broad wings moving in measured sequence and with the long neck and legs fully extended. The note is loud, clear, and trumpet-like, capable of being heard for immense distances. The Crane is for the most part a vegetarian, subsisting on grain of all kinds, grass, buds and leaves of water plants, acorns, and other seeds; its animal diet includes frogs, lizards, insects, and small fish. A flock of these birds, when feeding or resting, station sentinels to warn them of approaching danger. The Crane is rather an early breeder, the eggs being laid in the more southern localities in April, a month or so later in the far north. The huge bulky nest is placed upon the ground or in the shallow water in the least accessible part of the swamps and morasses; and as the birds are in the habit of returning annually to the same localities to breed, they probably pair for life. The nest, which is from two to five feet across, is made of heather, branches, sedges, and rushes, and lined with grass. The eggs are usually two, sometimes three in number, brownish or greenish buff in ground colour, blotched and spotted with reddish brown, pale brown, and grey. The female incubates them, and the young birds—clothed in buffish down—are able to run almost at once. The young and their parents remain in company until the migration period approaches, when these family parties unite into the large flocks which are so characteristic of the annual journeys of this magnificent bird. For the remainder of the season the Crane is gregarious, and the movements of these winter flocks are very regular.

The adult Crane has the general colour of the plumage slate-grey, shading into black on the quills; of these the innermost secondaries are very elongated, and fall in graceful plumes over the tail; from the eye along the side of the head and the sides of the upper neck is a white streak; the crown is bare of feathers, covered with scarlet warty skin; whilst the forehead and the lores are equally devoid of plumage, but covered with blackish bristles. The female closely resembles the male in colour, but the plumes are smaller. These are entirely wanting in the young, which have buffish margins to the feathers, and the bare parts of the head are clothed with plumage. The Crane stands nearly four feet high, and is from three to four feet in length.

THE GREAT BUSTARD

(OTIS TARDA)

THE knowledge that the magnificent Great Bustard was still a resident on English soil not sixty years ago is well calculated to awaken sad thoughts of regret in every reader who takes an interest in our native birds, and more especially in the preservation of disappearing or threatened species. There is no evidence at present to suggest that the Great Bustard ever was an inhabitant of Ireland, whilst in the remainder of the United Kingdom it seems to have been a local species confined to the champaign areas, or bare and open treeless districts. These were the Merse of Berwickshire, the open area of the Lothians, the wolds of Yorkshire and Lincolnshire, the warrens, heaths, and brecks of Norfolk, Suffolk, and Cambridgeshire, and the downs and naked uplands of Dorset, Wilts, Hants, Berks, Herts, and Sussex. Curiously enough, the earliest description of the Great Bustard in Britain is found in a work entitled *A History of Scotland*, written by Hector Boethius, and published in 1526. Since 1684 there appears to be no evidence that the Great Bustard dwelt in this area. Coming southwards, we find that the last Bustards disappeared from the Yorkshire wolds about 1826. Its final disappearance from Lincolnshire is not recorded, but Professor Newton states that it probably occurred about the same time. In Norfolk, where the bird appears to have lingered longest, the last two examples were killed in 1838. In Suffolk the Bustard ceased to exist in 1832; whilst the first ten years of the present century saw its extermination from Salisbury Plain in Wilts: similar remarks apply to Dorset. From its other English haunts it appears to have passed away without any record whatever, although we may mention that there is no evidence of indigenous birds occurring within the present century at all. It is somewhat difficult to account for the extermination of the Great Bustard in Britain by those causes which have been so disastrous in the case of other species. The planting of trees and the enclosure of land may have had some share in the extinction of the Bustard, but we are inclined more to attribute its disappearance to direct persecution from man. Much of the country formerly inhabited by this bird remains in a very similar condition to what it was when the Bustard roamed over it. That the bird can exist in well-cultivated areas is proved by its presence upon some of the most highly farmed land in the world in North Germany; and we can see no reason why this species should not be perfectly at home upon such places as the Norfolk "brecks" and the open land of the Wiltshire downs to-day, were reasonable protection afforded it. Another cause of its extinction was the introduction of the corn-drill and the horse-hoe, which led to the discovery of its nests, and of course to their destruction by ignorant farm labourers. The fact that the birds moult their quills so rapidly as for some time to be incapable of flight may also have helped in their extinction. Had the Bustard been carefully preserved during the breeding season, and only killed in reasonable numbers, and its capture with traps made illegal, there seems no reason why the bird should not have retained its place as an indigenous species down to the present time. Possibly the day may come again when the Great Bustard will be seen in the old haunts, for there is nothing to prevent its introduction being attended by success, if intelligently attempted, as was the case with the Capercaillie. Its sedentary habits are certainly in its favour. There can be little doubt that the indigenous Bustards were non-migratory. At the present time this bird is purely an abnormal winter wanderer to Britain, sometimes arriving in exceptional numbers, as during the winters of 1870-71, 1879-80, 1890-91.

A bird of the Bustard's wariness, gifted with long legs and ample wings, and frequenting the bare open country, is very well able to take care of itself under all ordinary circumstances. Notwithstanding this, even in some extra British localities the bird is not so numerous as formerly, especially in South Sweden (where, indeed, it is said to be extinct) and Denmark. If we admit the specific distinctness of *Otis dybowskii*, found in Siberia, China, and Japan, the range of the Great Bustard will include Central and Southern Europe and North-west Africa. It is said to visit Asia Minor, North Persia, Afghanistan, and North-west India. The favourite if not the exclusive haunts of the Great Bustard are treeless steppes and vast grain lands. It is more or less gregarious at all seasons, but most so in winter, when it unites into flocks of varying size, which roam the prairies in quest of food. A separation of the sexes into distinct flocks has been remarked at this season. During the summer immature birds remain in bands. In no part of its distribution are the migrations of this Bustard very pronounced. The bird is a very conspicuous one on the open steppes, especially before the grain or other herbage has grown sufficiently high to conceal it. Like most ground birds, it can make very good use of its legs, and if driven to flight soon passes out of danger with slow and deliberate beats of its ample wings. Its food is chiefly of a vegetable character,—grain, seeds, and the leaves and buds of plants,—but insects, mice, lizards, and frogs are also eaten. The usual note is a kind of grunt, and a hissing sound is produced by both sexes when alarmed or excited. This Bustard is said by some observers to be polygamous, but the balance of evidence seems to be in favour of monogamous habits, the birds pairing every spring. The greater scarcity of cock birds in England during the later years of the Bustard's occupation may have led to the assumption that several females lived under the protection of one male. The display of the cock Bustard in the pairing season is one of the most remarkable performances of its kind among birds. The nesting season is in May. The hen scrapes a hollow either on the open steppe or amongst the growing grain, lining it with a few bits of dry herbage. In this she usually lays two, and occasionally three eggs, olive green or olive brown in ground colour, spotted and blotched with reddish brown and grey. She alone appears to incubate them. If disturbed, she glides very quietly away, running for some distance before taking wing. But one brood is reared in the season.

The male Great Bustard has the head grey; the general colour of the upper parts is chestnut buff, barred with black; the primaries are black, the remainder of the wings white; the breast is banded with chestnut and grey; the remainder of the under parts is white. There is a tuft of long white bristly feathers or plumes on each side at the base of the bill. The female wants these accessory plumes, and the pectoral bands are absent. The male also possesses in some cases (possibly in very old birds) an air pouch or sac opening under the tongue, but its exact use is not yet fully ascertained. An old cock Great Bustard is from thirty-six to forty-three inches in length, and may weigh as much as thirty-five pounds; but the hen is considerably smaller, not much more than half that weight.

AVOCETS

THE AVOCET

(*RECURVIROSTRA AVOCETTA*)

HERE again we have a species which has been wantonly exterminated in Britain during the first quarter of the present century. The records of the persecution of this beautiful and curious bird are sad and exasperating in the extreme. Can it be believed that at the beginning of the nineteenth century the pretty, gentle, inoffensive Avocet was one of our commonest summer migrants to the fens and marshes of the eastern counties? Now—and for nearly eighty years, too—it is lost to us for ever; for no human efforts can restore it to the Fens again! Previous to that date there is evidence to show that its distribution in this country was much wider still. At the close of the eighteenth century the Avocet bred on Romney marshes, whilst there are earlier records of its presence in the Severn district and in Staffordshire. The last-known colony of

Avocets was situated at Salthouse in the Fen Country, but this was destroyed between the years 1822–25. It is recorded that the eggs were gathered from this colony to make puddings, and the poor birds destroyed for the sake of their feathers, which were used to make artificial flies! The drainage and enclosure of marsh land may have restricted the haunts of the Avocet; but experience has shown that a species is not readily extirpated by such means. To our lasting shame, we must attribute its extinction to the senseless persecution of the birds by man, and to the wholesale taking of their eggs, scientific collectors being to some extent responsible for the calamity. Parties of Avocets on migration still continue to visit East Anglia, especially in spring; but there is every reason to believe that these arrivals are not attempting to recolonise the deserted haunts, and whether the birds are captured or not is quite immaterial. We may rest assured that the bird as a breeding species is lost to us for all time. The fate of the British Avocets, however, might well serve as a warning in Denmark and Holland, where the bird is fast becoming rarer, and may eventually become extinct if measures for its protection are not taken in time.

THE AVOCET

Outside our limits the Avocet breeds on the southern coasts of the Baltic, on the Frisian Islands and the Dutch coast, as well as in the deltas of the Rhone and the Guadalquivir. Thence we trace it as a breeding species along the valley of the Danube and amongst the lagoons of the Black Sea. Still farther eastwards it is said to be resident in Palestine and Persia, and to breed in various parts of Central Asia, onwards to Dauria and Mongolia. To India and China it is a winter visitor; whilst in Africa it is more or less a resident throughout the continent, including Madagascar. The Avocet is a migratory bird, hence the impossibility of its ever being introduced into England by man. It arrives in flocks at its summer quarters in Europe during April and May, and quits them in September. Its favourite resorts are low sandy coasts, salt marshes, lagoons, and flat islands. Here it may be seen near the water, or wading in the shallows, or even swimming across deeper pools. It is not particularly shy, if wary, and will allow itself to be watched walking with graceful steps about the mud, or running over it if need be. A too close approach will cause it to soar into the air, where it flies with its long neck and legs outstretched and its black and white plumage giving it a curious aspect. At all seasons it is gregarious, and the effect produced by a large flock either standing on the mud or fluttering in the air is very singular and pleasing. The bird obtains its food by working its long slender upturned bill from side to side, and this food is composed chiefly of small worms, insects and larvæ, and tiny crustaceans, the captured morsel being swallowed with a toss of the head. The note of this species is a clear and softly uttered *tü-it*, heard most frequently when its breeding-places are disturbed by man.

In Western Europe the Avocet commences to breed in May. It nests in colonies, many pairs of birds occupying a small area of suitable ground. The nests are little more than hollows in the sand or mud, or amongst the short herbage, lined with a few bits of dry herbage. The three or four eggs are pale buff in ground colour, spotted and blotched with blackish brown and grey. Both parents incubate them, and but one brood is reared in the season.

The adult Avocet has the crown, the back of the neck, the primaries, scapulars, and a band across the wing from the shoulder to the end of the innermost secondaries black; the remainder of the plumage white. In the young the plumage is not so pure; the black has a brown cast, and many of the dark feathers have pale margins. The length of this bird is about sixteen inches.

THE BLACK-TAILED GODWIT

(*LIMOSA MELANURA*)

THE Black-tailed Godwit is another species which the exercise of a little ordinary care and common sense might have preserved. It seems almost incredible that in former days this bird was so common in East Anglia that it was regularly fattened for the table, and held in as much as or even greater estimation than the Woodcock is in our own. Its chief strongholds in Britain, so far as we possess any records, were in the fens of Lincolnshire and Norfolk and in the Isle of Ely. During the first quarter of the present century the Black-tailed Godwit bred commonly in the Fens; it ceased to do so about the year 1829, but a nest was found in Norfolk as recently as 1847. This Godwit still continues to pass over the British Islands in spring and autumn on its way to breeding-grounds farther north, but the stock of indigenous birds is gone, and we may safely conclude that the species will never nest with us again. This species furnishes another instance confirming the fact that the supply of our indigenous birds is not unlimited, and that if we unduly persecute them the time is sure to come when they will vanish from our avifauna. It is the breeding birds that should be jealously guarded; the winter visitors are not only better able to take care of themselves, but as a rule are much more numerous. So long as these individuals are not molested at their breeding-grounds in the Faroes, Iceland, and Scandinavia, Black-tailed Godwits will continue to visit us on passage. These may be met with locally on most of our coast-line, but are commonest on the mudflats of the east and south. Outside our limits the Black-tailed Godwit, in addition to the localities already given, breeds in Holland, Belgium, Denmark, Poland, Northern Germany, and Central and Southern Russia. Eastwards it is met with as a breeding species in Western Turkestan, and South-west Siberia as far as the valley of the Obb. In winter it is found on the Spanish coasts, throughout the basin of the Mediterranean, the coasts of the Red Sea, the basin of the Caspian, the shores of the Persian Gulf, and North-western India. In Asia, from the valley of the Yenisei eastwards, it is replaced by a closely allied form.

Lost as the Black-tailed Godwit is to British ornithologists, it may still be observed during the breeding season on the opposite coasts of the North Sea, in the marshy meadows of Holland, and in the fenlands of Jutland—proof, if proof were wanting, that the birds did not forsake their English haunts, but were ruthlessly driven from them. Drainage may have destroyed many an English breeding-place, but there are many others left where this bird could still have nested in peace. In Europe the spring migration of this Godwit begins as early as February, and continues through the two following months, those that cross the British Islands appearing in them in April and May. They are seen again on migration south in August and September, and in some places the passage lasts until October. This Godwit not only may be seen on tidal mudflats, but on salt marshes and the wet portions of moors. It is not exactly a shy bird, if a wary one, and Dr. Sharpe tells us that he has seen it standing complacently near the muddy dykes as the train rushed along between Rotterdam and Amsterdam; whilst on the Lincolnshire mud-flats we have repeatedly watched it running daintily about within easy gunshot. It flies well and rapidly, like all its allies, and frequently wades breast-high in the shallows. The food of this species consists of worms, insects and their larvæ, snails, and the seeds, buds, and roots of various plants. The call-note of this Godwit resembles the syllables *ty-ü-it*; whilst its cry, when alarmed at its breeding-grounds, is a loud and clear *tyü-tyü*. In Western Europe the breeding season of the Black-tailed Godwit is in May; occasionally eggs may be found late in April. Numbers of nests may be found within a small area of marsh. The nest is made upon the ground, in a tussock of sedge, or concealed amongst the herbage, and is merely a hollow, lined with a little dry grass or other vegetable refuse. The four eggs are olive brown in ground colour, spotted and blotched with darker olive brown, pale brown, and grey. But one brood is reared in the season.

In breeding or summer plumage the adult male Black-tailed Godwit has the head, neck, and breast reddish chestnut, marked with blackish brown on the crown and breast; the remainder of the upper parts (except the rump, which is white) are brown, more or less flecked and spotted with black; the wings are dark brown, with a conspicuous white bar across them; the tail is black, with a white base; the under parts below the breast are white, barred with brown on the flanks. The female is less showy than the male. In winter plumage the general colour of both sexes is greyish brown above and nearly white below the breast, which is marked with dusky streaks. In winter plumage the tail is ash grey, slightly marked at the base with white. The total length of the male of this Godwit is about sixteen inches.

THE BLACK TERN

(*STERNA NIGRA*)

WHETHER the extinction of this pretty Tern as a breeding species in England can be solely attributed to the drainage of fens and marsh lands is certainly doubtful, when we bear in mind how so many of our remaining species of Terns have been reduced in numbers by direct persecution and not the destruction of breeding haunts. The Lesser Tern is a sad example of this, and the greatest care will have to be exercised if we do not want to see it overtaken by the same lamentable fate. The Black Tern was formerly an abundant summer visitor to the fens and marshy lands of East Anglia; the drainage of these has curtailed its haunts, and in many places no doubt destroyed them. The last eggs of which any record has been kept appear to have been taken in 1858 in Norfolk. It is interesting to know that a few pairs of Black Terns appear annually in the districts the species frequented in such numbers years ago, and it is not improbable that these may be survivors of the old indigenous stock. They should be protected and encouraged, in the forlorn hope that the species may re-establish itself in this country. The fens and low grounds of East Anglia too long remained the happy hunting-ground of the fowler and the egg-gatherer, who have been permitted to destroy and take at any and every season, with the inevitable result that all true naturalists have now to deplore. In other parts of the British Islands the Black Tern can only be regarded as an accidental wanderer on abnormal migration. Outside our limits this Tern breeds as far north as Esthonia, thence southwards in the Baltic Provinces, Prussia, South Scandinavia, Denmark, Holland, France, the Iberian Peninsula, and eastwards through Central and Southern Europe to the Caspian. South of the Mediterranean it breeds in North Africa, excepting Egypt; whilst its Asiatic range includes South-western Siberia and Turkestan, east to the Altai. In winter this Tern is found as far south in Africa as the northern portion of the intertropical realm.

The Black Tern is a regular migrant to Western Europe, reaching its breeding quarters in May.

Its habits are very similar to those of allied birds. It spends most of its time in the air, gracefully flitting to and fro, dropping every now and then to the surface of the water to pick up some food. When on migration it may be seen flying along shore, but at other times it prefers to frequent fens, salt marshes, and swamps, and large sheets of water where the shallows are choked with reeds and rushes, and the alder trees form almost impenetrable thickets. At all times of the year it appears to be gregarious, and during summer lives in colonies of varying size to rear its young. The food of this Tern consists largely of insects, small fish, and other aquatic creatures, worms and grubs. The note is a shrill *crrick*, sometimes prolonged into *crree*. The nests of the Black Tern are made amongst the reeds in the shallow water, or on clumps of sedge and grass on the spongy ground of the surrounding marshes. They are bulky structures, like heaps of decaying vegetation, made of rotten reeds and sedges, and the hollow lined with dry grass. The eggs are three in number, and vary from buff to olive brown in ground colour, heavily marked with reddish brown, blackish brown, pale brown, and grey. Both parents assist in their incubation. When disturbed, the birds rise in fluttering crowds from the ground, with noisy cries of remonstrance, and continue to fly to and fro above their nests until the danger has passed. But one brood is reared in the season; and a movement south may be observed soon after the young can fly, the passage of this species extending from August to October.

The adult Black Tern in summer plumage has the head, neck, breast, and belly black; the under tail coverts white; the remainder of the plumage dark grey. In winter plumage the forehead, throat, and lores are white, and the under parts are more or less mottled with white. Young birds have the upper parts, especially on the head and back, mottled with brown. The length of this small Tern is about ten inches.

PLATE IV.

GREAT AUKS

THE GREAT AUK

(*ALCA IMPENNIS*)

THE species we have hitherto mentioned have become extinct in the British Islands only, their extermination being of a local character; but the present bird excites a wider melancholy interest, for there can be little doubt that it has ceased to exist altogether. Many erroneous opinions prevail not only respecting the geographical distribution of the Great Auk, but the cause of its extirpation. As most readers may know, the Great Auk was incapable of flight. The bird was nearly as big as an ordinary tame Goose, but closely resembled a Razorbill in general appearance, except that its short narrow wings were quite incapable of bearing it through the air. If useless for flight, these wings were used with marvellous power as oars, and the bird was a most accomplished swimmer and diver. This inability of the wings for flight was due to the abortive character of the bones of the forearm and hand, the humerus being proportionately as long as in the existing species of Auks, all of which are able to fly. As Mr. Lucas (one of the ablest historians of the Great Auk) points out, this modification of structure, however unfortunate it proved to its possessor, was correlated with the bird's aquatic habits; the resistance of water being much greater than that of air, a wing requiring less surface and more power than one formed exclusively for aerial locomotion would be best adapted for submarine flight.

Respecting the geographical distribution of the Great Auk, the impression widely prevails that the bird was an inhabitant of the Arctic regions; and more than one naturalist has suggested that the lost species may still be found in the Polar solitudes. Vain hope, with not a shred of evidence to support it! So far as is known, the Great Auk was confined to the North Atlantic, and there is no reliable evidence whatever that the bird ranged anywhere within the Arctic Circle.[1] On the eastern shores of the North Atlantic the bird ranged from Iceland to the Bay of Biscay, breeding certainly in the

[1] Professor Reinhardt says that there is doubt attaching to the locality of the specimen (now in the Copenhagen Museum) from Greenland, reputed to be from Fiskernäs, above the Arctic Circle.

Icelandic area, and possibly on the Faroes, the Orkneys, and some of the Norwegian islands.[1] There is little evidence to suggest that the Great Auk ever bred in any numbers, if at all, on St. Kilda, Martin's statements notwithstanding. On the western shores of the North Atlantic its range extended from Greenland to Virginia, but the actual breeding stations were few and far between. There can be no doubt that the grand headquarters of the Great Auk were on the American side of the Atlantic, and there the most important station of which we have any evidence at present was on Funk Island, off Newfoundland, although other breeding-places were possibly located along the coasts of Labrador and South Greenland. In European waters Iceland appears to have been the principal resort of the Great Auk, and from here most of the specimens of birds and eggs now in existence were obtained. Here the colony was located on several rocky islets situated some twenty-five miles to the south-west of the main island, the birds continuing to be fairly numerous, although harassed from time to time by collectors and others. But misfortune seems to have settled upon the Great Auk,

[1] Apparent remains of an egg have been discovered recently near Falsterbo, in South Sweden.

Nature herself hastening its doom in volcanic disturbances, which in March 1830 caused the principal breeding reef — the Geirfuglasker — to disappear beneath the waves, and compel the surviving birds to take up fresh quarters. Most of them appear to have selected the islet of Eldey—a very unfortunate choice, for this reef was situated much nearer to the main island, and was far more accessible to man. Here, within a period of fourteen years, every bird was killed, the last pair being captured early in June 1844, and forming the final record of the species in Europe. Coming now to British waters, we find it stated that two centuries ago the Great Auk was a regular summer visitor to St. Kilda, although, as previously stated, we doubt if the bird ever was established there in any numbers, the islets being for the most part very precipitous, and unsuited to its requirements. A bird, however, was caught there—in autumn be it remarked—as recently as 1821 or 1822; and we ourselves in 1884 were assured by an old inhabitant of the islands that a Great Auk was stoned to death as an "evil spirit" on Stack-an-Armin about half a century previous, he himself assisting in the massacre! In 1812, Bullock saw a Great Auk at Papa Westray in the Orkneys, and tried to shoot it without success, although the poor unfortunate was killed the following year, preserved, and sent to him. This specimen is now in the British Museum. The hen bird of this pair had been killed previous to Bullock's visit. One other British example was caught in a landing-net in Waterford harbour in May 1834, and is now preserved in Trinity College Museum, Dublin. Other evidence of the Great Auk's former existence in Ireland is presented in its remains found in some numbers on the coast of Antrim,[1] with those of the horse, dog, and wolf, and more recently in a "kitchen midden" in the county of Waterford. Remains of this bird have also been found in the superficial deposits in the Cleadon Hills in Durham, as well as at Oronsay and Caithness.

We now turn to the story of the Great Auk's extirpation in America,—a record of wanton cruelty and carnage that would be hard to beat,—"countless myriads of this flightless fowl," says Mr. Lucas, "hunted to the death with the murderous instincts and disregard for the morrow so characteristic of the white race." Although there is evidence to suggest that the bird was formerly abundant at

[1] *Irish Naturalist*, vol. v. p. 121: Proc. R. I. A. (3) iii. pp. 650-663.

Penguin Islands, off the southern coast of Newfoundland, Funk Island must have been the site of the most important colony. This latter locality was specially visited by Mr. Lucas in July 1887, on board the U.S. Fish Commission steamer *Grampus*, and from his intensely interesting accounts we will quote the following particulars.[1] Here, on the southern half of the island, "the Auk bred in peace, undisturbed by man, until that fateful day . . . when Cartier's crews inaugurated the slaughter, which only terminated with the existence of the Great Auk. The history of the Great Auk in America may be said to date from 1534, when, on May 21, two boats' crews from Cartier's vessels landed on Funk Island, and, as we are told, 'in lesse than halfe an hour we filled two boats full of them, as if they had been stones. So that besides them which we did eat fresh, every ship did powder and salt five or sixe barrels of them.' The Great Auk having thus been apprised of the advent of civilisation in the regular manner, continued to be utilised by all subsequent visitors. The French fishermen depended very largely on the Great Auks to supply them with provisions; passing ships touched at Funk Island for supplies; the early

[1] Report U.S. Nat. Mus. 1887-88; *op. cit.* 1889.

colonists barrelled them up for winter use, and the
great abundance of the birds was set forth among
other inducements to encourage emigration to
Newfoundland. The immense numbers of the
Auks may be inferred from the fact that they
withstood these drains for more than two centuries,
although laying but a single egg, and consequently
increasing but slowly under the most favourable
circumstances. Finally someone conceived the idea
of killing the Garefowl for their feathers, and this
sealed its fate. When and where the scheme
originated, and how long the slaughter lasted, we
know not, for the matter is rather one of general
report than of recorded fact, although in this
instance circumstantial evidence bears witness to
the truth of Cartwright's statement, that it was
customary for several crews of men to pass the
summer on Funk Island solely to slay the Great
Auks for their feathers. That the birds were slain
by millions, that their bodies were left to moulder
where they were killed, that stone pens were
erected, and that for some purpose frequent and
long-continued fires were built on Funk Island,
is indisputable." The final extinction of the Great
Auk in America was almost coincident with its
extirpation in Europe, the work of slaughter going

steadily on "until the last of the species had disappeared from the face of the earth, and the place to which it resorted for untold ages knew it no more." Mr. Lucas obtained the most ample evidence of the bird's former abundance. He tells us that "on the northerly slope a stroke of the hoe anywhere would bring to light at least a score of bones"; and again, "while many humeri were thrown aside while digging, the collection was found to contain over fourteen hundred specimens of this bone." The material brought back by him was estimated to be greater than that obtained by all other expeditions combined, and to include nearly two barrels of bones, from which ten or eleven skeletons of the Great Auk have been made up. Previous to the visit of Mr. Lucas to Funk Island, but two naturalists had explored the place. Stuvitz went there in 1841, and discovered some bones; Professor Milne visited the island in 1874, and after an hour's work brought away bones belonging to some fifty birds and the inner linings of several eggs; whilst nine years previous to the latter naturalist's visit, an expedition sent out for guano procured three "mummies" or dried bodies of the Great Auk.

The extinction of this noble bird is all the more

to be regretted when we bear in mind that it was absolutely avoidable and unnecessary, and was in no remote way due to those economic and industrial changes which have deprived so many other species of a home. Here in the present case we find no invasion by civilisation of favourite haunts, no destruction for the sake of improvement of time-honoured breeding-grounds, no increase of population to exterminate timid creatures, but simply a cruel and wanton massacre of poor helpless and defenceless birds for the sake of commercial greed and gain that really could have had very little value. The extermination that went on in Iceland in an era of greater intellectual activity has even less to defend it; for there the latest survivors of the Great Auk were captured to supply various scientific institutions in Europe, so that literally its extirpation was countenanced and approved by and was undertaken in the name of Science! There was no reason whatever why the Great Auk should not have survived and even flourished in our own day. It is true the bird was comparatively helpless, but its inability to escape from enemies only prevailed during the nesting season, when the poor bird was engaged in duties that should have ensured for it immunity from harm. At all other times it was practically safe in its natural element the sea. Regrets are useless now; and when the few relics that are in existence have mouldered away, the Great Auk will fade from our memories, live but as a tradition, and finally perhaps as a legend or a myth!

Notwithstanding the former abundance of the Great Auk, and its comparatively recent final disappearance, but very little indeed is known respecting its habits. These, there can be little doubt, were very similar to those of its surviving allies, especially of the Razorbill, its nearest living relation. We know that it was an accomplished diver, we also know that it lived on fish; but of its notes, its nesting habits, its migrations, and the like, history is silent, and records are wanting. The breeding-places of this species were flat rocks that sloped gently to the sea, and the single egg was, it is presumed, laid nestless on the ground. This egg runs through similar variations to those of the Razorbill, but is, of course, double the size. The number of eggs at present known to exist is seventy-one. There are also seventy-seven skins of the Great Auk in various collections, together with many more or less complete skeletons and large numbers of odd bones.

THE GREAT AUK

The Great Auk has the general colour of the upper parts, including the wings, black; the secondaries are tipped with white; the tail is black; the throat is black; the remainder of the under parts white, as is also a large patch on each side of the face between the base of the bill and the eye. Bill similar to that of the Razorbill, but the white grooves not quite so conspicuous. In winter the throat became white, as in the Razorbill. The length of the Great Auk was about twenty-five inches.

VANISHING BRITISH BIRDS

THE BEARDED TITMOUSE

(*PANURUS BIARMICUS*)

THE birds we now come to deal with are fortunately still indigenous to the British Islands, although they are present in sadly diminished numbers, and are all more or less threatened with extinction in our area unless efforts are taken to preserve them and senseless persecution is relaxed. Our first species is the Bearded Titmouse, although why it should be called a "Titmouse" is hard to say; for its habits, characteristics, and organisation show little or no direct relationship with the group, and its true affinities remain yet to be discovered. This charming little bird is not only one of the prettiest, but one of the most interesting of our native

BEARDED TITS

species. It is also one of the most local. We have evidence to show that formerly the Bearded Titmouse occupied a much wider area in England than is now the case. This area included Lincolnshire, Cambridgeshire, Huntingdonshire, Norfolk, Suffolk, Essex, Kent, Sussex (possibly Hants), Dorset, and Devonshire. Probably it also occupied suitable districts in the valley of the Thames, even as far as Gloucestershire. At the present day this range is sadly curtailed, and only includes the counties of Devon, Suffolk (possibly), and Norfolk. When we come to investigate the causes of such rapid and wholesale restriction of area, we find it directly attributable to the destruction by drainage and enclosure of haunts, and to the direct persecution of man. We know that vast areas where this bird formerly dwelt have been improved away; the forests of reeds and the wet lands have vanished, and with them have gone the Bearded Titmouse. But this can only explain part of the extinction of the species. There are many wide areas left that the bird was known at one time to inhabit, but which are now apparently deserted, and these haunts have been decimated in the interests of collectors. Not only have marsh men taken every nest they could find, but the parent birds have been captured too. Here again we find the supply of birds limited and unable to fill the demand. Not only so: the Bearded Titmouse is a resident species, strictly confined to its native reed beds, so that when the British stock becomes exhausted the bird will pass out of our fauna completely, as so many other interesting forms have already done. We are heartily glad to hear that in some districts measures are being taken for the better protection of the Bearded Titmouse. We trust that these may prove successful, be more generally applied, and strictly enforced; for there is evidence to show that the bird in some districts especially is rapidly diminishing in numbers. We appeal to the owners of the reed beds frequented by this species to preserve it from extinction, and hope that local Natural History Societies will exert their widespread influence in the good cause.

Beyond the British area the Bearded Titmouse has a most extensive range, being found over a great part of Europe and Asia, at least as far east as North-eastern Thibet. We find it an inhabitant of the reed beds of Holland, Pomerania, and Hungary, in France in the marshes of the Rhone and Narbonne, in Spain, eastwards to Italy, South Russia, Turkestan, and South Siberia. To Holland

and Germany it is said to be a summer visitor only, but further information is desired. Examples of this species become paler towards the eastern limits of its distribution, and Central Asian birds were described by Bonaparte as *Panurus sibiricus*. As birds almost if not quite as pale may be met with in the extreme western areas, this form can only be regarded as sub-specifically distinct. We have no record of the Bearded Titmouse south of the Mediterranean or north of Pomerania, whilst it is extremely rare and local in the Levant.

The favourite, we might almost say the exclusive haunts of the Bearded Titmouse are reed beds. In England these are few and far between nowadays. It is a somewhat secretive species, skulking amongst the reeds and sedges when too closely approached, although sometimes seen flitting across the open waterways in an uncertain, undulatory manner, or clinging to some tall bending stem. During autumn and winter the Bearded Titmouse, or "Reed Pheasant," as it is locally termed in the Broad district, lives in flocks and parties of varying size, which roam about the reed forests in quest of food; but in spring and summer it is met with in pairs alone. Seebohm, who specially visited the Broads to observe the habits of this bird, describes its note as a musical *ping*, its alarm-note a harsh, Whitethroat-like *chir-r-r-r*, and its cry of distress a plaintive *ee-ar-ee-ar*. The food of the Bearded Titmouse is composed in summer of insects and tiny molluscs; in winter, of the seeds of the reed and other plants. Of the pairing habits of this species nothing definite is known. Its nesting season begins in April, and is prolonged until July, two broods being reared in the year. The nest is generally made beneath the shelter of a tuft of sedge or other coarse aquatic herbage, well concealed by overhanging vegetation. It is cup-shaped, rather deep, and made externally of dry grass, dead leaves, bits of reed, and scraps of withered aquatic plants; internally of finer grass and the flowers of the reed. The eggs are from five to seven, creamy white in ground colour, freckled with irregular lines and specks of dark brown. From these few particulars it may be remarked that the Bearded Titmouse is somewhat prolific, and we believe would hold its ground and steadily increase if reasonable protection were afforded it.

The adult male Bearded Tit has the head delicate lavender grey; the lores and a tuft of drooping, moustache-like feathers on either side of the gape are black; the general colour of the upper parts is rufous brown, shading into pinkish brown on the upper tail coverts; and the tail feathers are tipped with dull white, the outermost feathers with a margin also of the same tint; the wings are dark brown, the primaries with white margins and tips, the secondaries with rich rufous ones, the scapulars rusty white; the lesser wing coverts greyish brown, the greater ones black, both tipped with rufous, and the latter margined with the same. The throat and breast are grey, with a rosy flush; the centre of the belly is pale buff; the flanks are rufous brown, the under tail coverts black. The bill and irides are yellow. The female is not so brilliantly coloured; the black on the head and the moustache are wanting, and the under tail coverts are rufous. Young birds resemble the female in colour, and have the crown and back streaked with black. The total length of this bird is about six inches.

THE ST. KILDA WREN

(*TROGLODYTES HIRTENSIS*)

PERHAPS we may be forgiven for taking an exceptional interest in the fate of this bird; for we had the pleasure of ascertaining that it differed in certain respects from the Wren found in other parts of the British Islands. In 1884, when we brought the first known specimen from St. Kilda, the bird was common enough on all the islands of the group, and its cheery song could be heard everywhere. No sooner, however, was its specific distinctness pointed out by Seebohm in the *Zoologist* and by ourselves in the *Ibis*, than it became a coveted object by collectors of British birds and eggs, and specimens of both were eagerly sought. The natives of St. Kilda, urged on by the greed of gain, were not slow to take advantage of such an opportunity for making money, and the species has suffered sorely in consequence. That it will ultimately become as extinct as the Great Auk which once frequented these Atlantic isles, is certain unless strong measures are taken by the proprietor of the islands for its protection. Many pairs, there can be little doubt, still frequent the uninhabited portion of the group; so that, if proper steps be taken, we may succeed in saving from extinction so interesting an example of an island race of the familiar Wren. Our discovery seems always clouded with the exterminating results that have followed it, and when we hear of the poor bird's decimation we feel that, in the interests of science, it would have been better had we remained silent. It is sad to think that the publication of such knowledge resulted in absolutely threatening the extirpation of the St. Kilda Wren, and that by calling attention to its differences we have been the unintentional means of its being sacrificed to the greed and selfishness of collectors. We appeal to British naturalists to save this island form of the Common Wren from extirpation, threatened as it is by no other danger than that arising from the mania for possessing its eggs and its skin. The wholesale collecting of specimens by St. Kildans, and by tourists that visit the islands in summer, when the bird is breeding, must be sternly forbidden if the Wren is to be saved.

The following account of the habits of the St. Kilda Wren was the first published, and was contributed by us to the *Ibis*:—" I had not been on St. Kilda long before the little bird arrested my attention, as it flew from rock to rock, or glided in and out of the crevices of the walls. It differs very little in its habits from its congener; only, instead of hopping restlessly and incessantly about brushwood, it has to content itself with boulders and walls. It was in full song, and its voice seemed to me louder and more powerful than that of the Common Wren. I often saw it within a few feet of the sea, hopping about the rocks on the beach; and a pair had made their nest in the wall below the manse, not thirty yards from the waves. I also saw it frequently on the tops of the hills and in many parts of the cliffs. It was especially common on Doon, and its cheery little song sounded from all parts of the rocks. As there are no bushes nor trees on St. Kilda (except those the microscopic eye of a botanist might discover), the Wren takes to the luxuriant grass, sorrel, and other herbage growing on the cliffs, and picks its insect food from them. It also catches spiders and the larvæ of different insects in the nooks and crannies which it is incessantly exploring. It is a pert, active little bird, by no means shy; and I used to watch a pair that were feeding their young in a nest not six yards from our door. Its breeding season must commence early in May, for the young were three parts grown by the beginning of June. It makes its nest either in one of the numerous 'cleats,' or in a crevice of a wall, or under an overhanging bank. The nest is exactly similar to that of the Common Wren, and abundantly lined with feathers. The eggs are six in number, perceptibly larger and more heavily marked than those of the Common Wren, but otherwise closely resembling them. I found the birds remarkably tame at the nest, going in and out as I stood watching them. Probably two broods are reared in the season."

The St. Kilda Wren somewhat closely resembles the Common Wren in appearance, but is larger, more distinctly barred on the upper parts, and has much stouter feet. The variations of plumage due to age, sex, and season are not known to differ in any important respect from those of the allied forms.

THE HOOPOE

(*UPUPA EPOPS*)

FOR two centuries or more this beautiful and curious bird has been known to visit the British Islands in spring to breed. There can be little or no doubt that in the olden times the Hoopoe was commoner and more widely dispersed than it is now, and that, like so many other interesting species, it has been well-nigh exterminated for the sake of its beauty or novel and curious appearance. The British stock of Hoopoes, however, does not yet seem to be quite exhausted, and we may still regard the bird as a regular spring migrant to the southern counties of England. We must, however, bear in mind that the constant persecution which the species suffers in our islands, the failure to rear offspring in them, must sooner or later end in the complete extirpation of the Hoopoe as a British bird. We doubt not that careful preservation for a few years would end in complete reinstation of the species, and stock our southern English counties with Hoopoes, which might eventually spread northwards; for the bird breeds on the Continent as far north as South Sweden and Denmark. The Hoopoe has absolutely been known to breed in Devon, Dorset, Wilts, Hants, Surrey, Sussex, and Kent. To other parts of the British Islands the Hoopoe, at present, can only be regarded as an abnormal migrant, although the evidence seems to show that in the south of Ireland the bird may be a normal spring migrant, only requiring a little judicious preservation to establish it as a regular breeding species. When we read that no less than seven of these charming birds fell to one gun in a week on a single Sussex estate, or that a certain spot in Kent, apparently in the direct line of migration, is notorious for its butchery of Hoopoes, we may reasonably protest and demand that such slaughter shall cease. We fear that legislation is powerless without public co-operation, and this surely need not be asked in vain in such a compassionate age as ours! Unfortunately, the Hoopoe is a very conspicuous bird, and also a confiding and unsuspicious one, easily approached and shot.

Outside our limits the Hoopoe has an exceedingly

wide distribution, extending from Denmark in the west to Japan in the east, and from South Sweden and Central Asia to tropical Africa, India, and South China. It is found in suitable localities throughout Central and Southern Europe, and is very common in all Mediterranean countries and the Canary and adjoining islands. The Hoopoe is a summer visitor to Europe, arriving from February onwards, and usually reaching the south of England in April. The return passage takes place during August, September, and October. Its favourite haunts in Europe are well-cultivated districts, the fields on the borders of woods, and the more open parts of forest lands; but in Algeria I found it dwelling on the bare hillsides, as well as in the palm-studded oases, where the Arabs let it run about outside their houses and tents without any attempt to molest it. It is an active, graceful little bird, and may often be watched strutting about in quest of food on the dunghill or newly-tilled land. In its summer quarters in Europe it shows much more partiality for cover than in its winter ones in Africa. Its note is a hollow, deep, and full-sounding *poo-poo-poo*, or *hoop-hoop-hoop*, capable of being heard for a long distance. Its principal food consists of insects, worms, and grubs.

The flight of this species is undulating, like that of a Wagtail or a Woodpecker, and when in the air the bird becomes even more conspicuous, as its particoloured plumage is fully displayed. The Hoopoe probably pairs for life, and appears to return to a certain spot annually to breed. The nest is usually made in a hole in a rock, or a wall or tree, but is never excavated by the birds. It is merely a small collection of dry grass, straws, or roots, more or less mixed with offensive matter of some kind, and causing a fearful stench, which becomes even worse as the droppings of old and young accumulate. The eggs are from five to seven in number, and vary from pale greenish blue to lavender grey and buff, and are without markings. The shell is coarse and full of minute hollows. The female alone incubates them, and the male is said to feed her assiduously during the task. But one brood is reared in the year, and after the breeding season the birds often gather into small flocks and family parties for the winter.

The adult Hoopoe has the head decorated with a very handsome erectile crest formed of broad feathers, buffish chestnut tipped with black, and with a narrow subterminal bar of white; the remainder of the upper half of the bird is chestnut buff, darkest on the back, and suffused with pink on the breast; the lower half of the bird is curiously pied. The lower back, scapulars, and innermost secondaries are buff, marked with black; the flight feathers are black, broadly barred with white; the rump is white; the tail is black, with a broad white band across the centre and spreading towards the tips of the outer feathers; the belly and under tail coverts are white; the flanks are streaked with dark brown. The female is a trifle smaller and duller, as are also the young. The total length of this bird is about ten inches.

THE OSPREY

(*PANDION HALIÆETUS*)

IT is rather a remarkable fact that not a single species of raptorial bird has been exterminated in the British area within the historical period. Of all species, the birds of prey have been subject to the greatest amount of steady and persistent persecution, and yet they have managed to survive. Many of them, however, once common and widely distributed, have become excessively local; others that formerly bred in England now only survive in the wilder areas of Scotland, Ireland, and Wales. Generally speaking, birds of prey are thinly distributed, not collected in certain spots like more gregarious species; and there can be little doubt that to this fact they owe their survival. There is a great falling off in the number of most raptorial species, owing to the systematic trapping, poisoning, and shooting which has gone on ever since the preserving of game has been so widely practised; their eggs have been persistently sought and taken, their young destroyed; and yet, in spite of all, not a single indigenous species has succumbed absolutely, although it may have been extirpated in many districts. We heartily hope that more sensible opinions will shortly prevail concerning the economic value of many of these raptorial species, and that, duly protected, they will become more numerous, to the benefit of the agriculturist.

The Osprey, if tradition is to be believed, once bred upon the southern coast of England; whilst a hundred years ago, upon the authority of Heysham, it bred in the Lake District, near Ullswater. Forty years ago two eyries were known to exist in Galloway; but at the present time we believe the sole stronghold of the Osprey is in the Highlands—fortunately in districts where the bird is protected and its haunts kept secret. Perhaps in time this privacy may not be necessary, but nowadays the Osprey retains its place in our fauna with such a slender hold that naturalists cannot be too careful in guarding its last retreats from the intrusion of the bribing collector of rare birds and eggs. Certainly, so far as Scotland is concerned, we cannot attribute the present rarity of the Osprey to the destruction of its haunts, and we are compelled to assign the direct persecution by man as the reason of its untimely disappearance. Not only has the bird been robbed of its eggs and young and shot in Scotland, but numbers continue to fall victims to the gunner in more southern districts whilst performing their annual migrations. As a visitor on passage, and especially in autumn, the Osprey is fairly well known in various parts of the British area, both near inland waters as well as along the rivers and coasts, especially of the eastern and southern counties. Unfortunately, too many of these Ospreys are killed, and we would forbid the shooting of this species within British limits altogether. To Ireland the Osprey is an abnormal migrant only. Beyond the British area the Osprey has an exceedingly wide distribution, breeding in all suitable localities throughout Europe, Asia, North America, and Australia, although the birds inhabiting the latter area present differences which may have a sub-specific value. In Europe it may be met with, breeding from Lapland to the Iberian Peninsula, and from North Russia to the Caspian; whilst south of the Mediterranean it nests in many parts of North Africa, from the Canaries (where it is said to be a resident) to the Red Sea. It is a summer migrant in the northern portions of its range, and a winter one in many of the southern limits.

Of all the raptorial birds the Osprey is the most aquatic in its habits, and its haunts are always more or less close to waters well stocked with fish. In our islands the favourite haunts of this bird are the wild mountain deer forests, the hill-surrounded lochs and quiet lakes studded with islands, on many of which some ancient fortress or ruined tower tells of warlike deeds of the long ago. Although many of these secluded Highland waters literally teem with fish, the Osprey is far too rare a bird to be seen near them in any numbers, an isolated pair here and there being all; but in North America, where the species is a much more abundant one, large colonies of these birds may frequently be met with. The Osprey reaches its breeding-grounds in Scotland in April or May. We have had few opportunities of studying this bird in a wild state; but we can vividly recall our first sight of the Osprey in its native land, close to the head-waters of Loch Carron. The bird was about thirty feet above the water, passing along, hovering every now and then with quivering wings, alternated with rapid beats, as is so often the way of our better-known Kestrel. Finally we watched it poise for a moment and drop down, Gannet-like, into the water, the noise as it struck the surface being distinctly audible from the shore. The bird rose again in a few seconds, and slowly retired to a distant clump of trees, but whether it had caught a fish or not we were unable to determine. In its search for prey the Osprey is very Gull-like, but of course seizes its food with its feet, and not with its bill. This food is composed of fish, such as trout, roach, bream, shad, flounders, etc. These are always captured with the feet, the soles of which are very rough, and the long claws exceptionally sharp. The note of the Osprey has been described as *kai-kai-kai*, and when alarmed the bird is said to utter a harsh scream.

The Osprey most probably pairs for life, and returns to one locality to breed year after year. In the Highlands nowadays the nest is generally made on the broad flat top of a pine tree, but formerly it was as frequently placed on ruins or rocks on islands. The nest is an immense pile of sticks, the accumulation of years, perhaps as much as four feet high and as many broad, intermixed with turf and lined with green grass. Sometimes several nests are made in one locality, and used in turn. The two or three eggs are very handsome, white or pale buff in ground colour, heavily blotched and spotted—sometimes so densely as to conceal the pale ground—with rich reddish brown, orange brown, and grey. But one brood is reared in the season. It is said that the Osprey will savagely attack an intruder of its nest. Professor Newton says that men and boys have had their heads gashed with the sharp claws of the enraged parent bird. In North America as many as three hundred nests have been found in trees close together.

The Osprey has the head and nape white, streaked with brown, some of the feathers being elongated. The general colour of the rest of the upper parts is dark brown, occasionally shot with purple; the under parts are white, except a band of brown spots across the breast. The female is similar to the male in colour, but she is slightly larger, and the head and breast are more marked with brown. Young birds resemble the female in colour. The total length of the Osprey is about twenty-three inches.

PLATE VI.

THE KITE

THE KITE

(*MILVUS REGALIS*)

TO realise the amount of persecution that raptorial birds have suffered in the British Islands, we have only to recall the days when the present species was spoken of by old writers on Natural History as one of the most abundant and widely distributed of our indigenous birds. Old records inform us that four or five hundred years ago the Kite literally swarmed in London, and that the bird was actually protected by law within the precincts of the City! Indeed, the Kite was formerly held in esteem for its good offices as a scavenger. We have Belon's testimony that he found the Kite scarcely less numerous in London than in Cairo, and that it cleared the streets and the river of garbage and refuse. Further, the many allusions, both poetical and otherwise, to the Kite in our literature eloquently speaks to the bird's former abundance. Even less than a hundred years ago the Kite was by no means a rare bird; from what we can gather, the bird was a by no means unfamiliar object in the rural scenes of England, floating high in air above the fields and woods, indulging in those magnificent flights which justly gained for it the name of Glead or Glide Hawk. Turner tells us that the Kite even snatched food from the hands of children in our towns—a fact which proves how little subject to persecution the bird must have been, and how bold and impudent it became in consequence. But as the present century sped its course, the preservation of game became more general and more strict, firearms were improved, and the Kite must very rapidly have decreased in numbers. The decay of hawking must also have had an evil effect; for the Kite was a prized quarry, and preserved accordingly. As the bird became rarer, the collector of eggs and skins must also be included as an exterminating agent; whilst in Scotland the bird was being ruthlessly killed for the sake of its tail feathers, which were highly prized for the purpose of making salmon flies. The result of all this persecution is that the Kite has become one of the very rarest of our indigenous birds. It is still left to us, still lingers in one or two localities, but there can be no doubt that the species will become extinct in our area unless great care be exercised. There can be little doubt that the Kite breeds in few parts of England at the present time, one of the last recorded nests being taken in Lincolnshire twenty-seven years ago. It, however, still continues to breed in Wales, and in one or two localities in Scotland; whilst Professor Newton states that it nests at the present time in certain woods in Huntingdonshire and Lincolnshire, and in the Western Midlands. The Kite never seems to have been indigenous to Ireland; and nowadays the bird is only occasionally seen in most parts of England—individuals apparently on migration, and probably far out of their usual course. Outside the British area the Kite breeds in most parts of Europe, resident in the central and southern districts and migratory in the extreme north. In Scandinavia it reaches as far north as lat. 61°, and in Russia certainly nests as high as Archangel. It is best known in North-west Africa as a winter visitant, but a few remain to breed; and it is also found in the outlying Atlantic islands. In North-east Africa it appears to be absent, although it is a common winter visitor to Palestine, where it also breeds.

Its eastern limits appear to be the valley of the Don.

This fine bird may be easily recognised even upon the wing by its deeply forked tail. One of the most remarkable characteristics of the Kite is its singularly graceful and long-sustained flight. When the bird was far more common than it is now, the English naturalist could stand and watch its amazing powers of wing by the hour together. Upon one occasion only in the British Islands have we had the good fortune to witness the once familiar sight of a soaring Kite. We watched the bird rise from a grove of trees and mount upwards and upwards in spacious circles with wings and tail outspread, the highest air being gained with scarcely an apparent effort, whilst all the time the beautiful creature was bearing away from us until we literally lost it in the clouds. Falcons were formerly flown at the Kite, and it was no uncommon thing for the latter to keep the air of its pursuer until both were lost to view. In many of its habits the Kite closely resembles the Buzzard; it is ordinarily a sluggish bird, without any of the splendid dash which characterises the movements of the Hawks and Falcons. From this we may infer that the food of the Kite consists principally of young and

THE KITE

weakly birds and small animals, insects, carrion and offal, and occasionally fish. The mewing note of the Kite in this country is rarely heard except during the breeding season.

We may here state that the Kite is doubtless a resident in our islands, such few as breed here. Its favourite resorts for nesting purposes are woods. The nest in Britain seems always to be made in a tree, but in North Africa a ledge of rock is often selected. A pine or fir tree is preferred. It is placed sometimes amongst the more slender branches at the top of the tree, but more frequently in a crotch lower down and close to the trunk. Externally this nest is made of sticks, often festooned with rags, waste paper, and such-like rubbish; internally it is lined with moss, wool, bones, fur, hair, rags, and even twine. The eggs are usually three, but sometimes two or even four, in number, pale bluish green or almost white in ground colour, blotched, spotted, and streaked with dark reddish brown, paler brown, and grey. The Kite is single-brooded, and the eggs are laid in May.

The Kite has the general colour of the upper parts reddish brown, the feathers with paler edges; those on the crown and neck somewhat elongated.

greyish white streaked with brown. The under parts are rufous brown streaked with dark brown; the tail reddish brown, with several dark bars. The female is larger than the male, but is very similar in colour. The length of the Kite is about twenty-two inches.

THE COMMON BUZZARD

(*BUTEO VULGARIS*)

A BIRD whose name bears such an epithet before it seems out of place in a volume like the present; but, alas! "common" it can no longer be called, it is only such in name, and is yearly becoming rarer, probably becoming extinct if persecution be not relaxed. Formerly the Common Buzzard was fully entitled to its name, being pretty generally distributed throughout the British Islands. Half a century ago this bird could not be called rare; five-and-twenty years ago we took a nest ourselves in Derbyshire. The rarity of this species now is largely due to the gamekeeper; as one of these worthies assured us many years ago, a "Hawk as big as a coal-basket must do a tremendous lot of harm." It has been ruthlessly shot down accordingly, and that, too, without any justifiable cause; for the Buzzard is by no means the enemy to game that sportsmen imagine, its food consisting of many creatures eminently more destructive. Of its harmlessness we hope to convince the reader in our account of its general habits. Collectors, too, are responsible to a great extent for the Buzzard's present rarity. British-taken eggs have long been a craze, and the high prices which these have commanded have tempted woodmen and keepers to seek for and rob many a nest which might otherwise have been left alone. If we are to retain the Common Buzzard as a British species, this persecution must cease; and we trust that County Councils, agriculturists, and big landowners will in their own interests save the bird from the extinction which threatens it. Nowadays the Common Buzzard still breeds on some of the Welsh cliffs and in the larger woods of the Principality, as it also does in a few of the wilder wooded districts of England. In Scotland we are glad to say it has not been so sorely persecuted, and still breeds in many secluded spots; whilst in Ireland, although far less common than it used to be, it is known to nest here and there in localities which are best kept secret. The range of the Common Buzzard outside the British Islands is a somewhat restricted one. The bird is generally distributed over Western Europe from about lat. 60° north in Scandinavia southwards to the Mediterranean, the valley of the Danube, the Black Sea, and the Volga Delta. In North Africa it is replaced by nearly allied forms, but the typical species seems to be the one that breeds on the Canary Islands. In the extreme north it is a migrant, but in the southern areas it is resident.

The Common Buzzard is a resident in the British Islands—that is to say, the indigenous individuals. In its habits it is somewhat sluggish, wanting the nimble movements and the impetuous dash that characterise the Falcons and the Hawks. Its flight is usually slow and laboured, the wings beaten deliberately; but on occasion the bird displays some exceptionally fine aerial movements, as, for instance, when it ascends in a spiral manner to a vast height, usually in the breeding season, and above the woods that contain its nest. It hunts for food in a very patient manner, often sitting on a fence or bare limb of a tree waiting for some small animal to appear, which it drops down upon and secures. Its food consists very largely of field mice (a nest visited by Seebohm contained no less than eleven field mice), frogs, small snakes, and rarely birds. Indeed, this Buzzard is of as much service to the agriculturist as the Owl, and should be protected and encouraged accordingly. Seebohm describes the note of this bird as a melancholy *pe-e-i-o-oo*. In our islands the breeding season of the Common Buzzard is in April and May. Its British breeding haunts are not only in large woods, but on maritime cliffs, and as it returns to a locality year by year to rear its young, it may not improbably pair for life. The nest is either made in a tree or on a ledge of some cliff; when in the latter situation, frequently made amongst ivy or under the shelter of a bush. It is large, flat, and made externally of sticks, lined with finer twigs, a scrap or two of wool, and quantities of green leaves—the latter apparently being renewed from time to time. The eggs are from two to four in number, usually three, and vary from white or pale buff to pale bluish green in ground colour, blotched, splashed, and spotted with reddish brown, paler brown, and grey. The female performs most of the duties of incubation, and when flushed from her charge sometimes circles round the spot uttering a monotonous note. This species is single-brooded.

The Common Buzzard is a species that presents considerable variation in the colour of its plumage, and a description of these would take up far more

THE COMMON BUZZARD

space than can be allotted. Speaking generally, the bird has the general colour of the upper parts brown, the scapulars and wing coverts with paler margins; the head and nape are more or less streaked with white. The under parts are whitish, shading into brown on the breast, flanks, and thighs; the primary quills are brown, with darker bars; the secondaries are paler; and the inner webs of all are white for two-thirds of their length. The tail is brown, crossed with about a dozen bars of darker brown. The female closely resembles the male in colour; and the young are said always to be paler than adults. The length of this Buzzard is about twenty-two inches.

THE GOLDEN EAGLE

(*AQUILA CHRYSÆTUS*)

IT is certainly matter for surprise that a bird as big as the Golden Eagle has managed to retain its place in our avifauna, and we must attribute the circumstance to the inaccessible character of those remote haunts it now affects. Little more than two hundred years ago the Golden Eagle bred in Derbyshire and Wales. Willughby says that this bird in his day bred on the cliffs of Snowdon, and he actually describes an eyrie in Derbyshire in 1668. Wallis, a century later, publishes the information that it bred on the Cheviots; whilst Jardine, in 1838, is able to give the cliffs of Westmorland and Cumberland as recently its breeding-place. Probably the bird formerly bred in many parts of England and Wales; but persecution has done its work, and we shall never see the Golden Eagle an inhabitant of the Lowland shires again. In the Lowlands of Scotland the

PLATE VII.

THE GOLDEN EAGLE

bird still lingered as a breeding species to somewhere about the year 1855; now the Highlands are its only resort. In Ireland the bird has sadly decreased in numbers, and its principal breeding-places are, according to Mr. R. J. Ussher, a few places in Mayo, Donegal, Galway, and Kerry. Returning to Scotland, the Golden Eagle breeds here and there in the Hebrides and the Western and Northern Highlands. It is most satisfactory to know that this splendid bird maintains its ground, and that in some districts it is actually increasing, thanks to the protection which has been given it by certain landowners, to whom all naturalists must feel more than grateful. Collectors, especially oologists, are responsible for the extermination of a good many Golden Eagles; and when we hear of the tempting prices which are offered to shepherds and others, the only wonder is that the bird exists at all! We are convinced that saner opinions are spreading, and we have every confidence that the bird for the present at all events is safe. Let the splendid bird be guarded as national property, for its presence in the Highlands is an ornament that cannot be spared. Beyond British limits the range of the Golden Eagle is a wide one, exceeded by few other species.

It is generally distributed throughout Europe, Asia, North Africa, and North America. We find it dispersed in Europe from Lapland to Spain, and across Asia to Kamtschatka and Japan, southwards to the Himalayas; whilst in the New World it is found from the Arctic regions to the States, although absent from Greenland.

The Golden Eagle is a sedentary species in the British Islands, although one that wanders about a good deal during the non-breeding season. Its favourite haunts are mountains and glens and the secluded fastnesses of deer forests. It may be frequently met with in marine districts as well as inland ones. The most striking feature in this bird is its magnificent motions in the air. Its flight, so powerful and so long-sustained, ever calls forth our warmest admiration; and to watch the big broad-winged bird soaring in majestic curves high up in the blue sky is a sight that impresses itself on the memory for ever. Time after time in the Highlands has it been our good fortune thus to watch the Golden Eagle on the wing, sailing and soaring among the mountain tops, and occasionally swooping earthwards with erected wings in all the majesty of freedom. Except in the air, this Eagle is a somewhat sluggish bird, fond of sitting motionless on the rocks, basking in the sun and digesting its food. The Eagle, we think wrongfully, is accused of playing sad havoc amongst game; but its favourite food, as was long ago remarked by the observant Charles St. John, and as we have repeatedly ascertained, is the mountain hare, and failing this, on carrion, especially on dead sheep, which are common enough on the vast farms. Its love of carrion often leads it into traps, and brings death by poison; weakly lambs and deer calves, together with wounded or diseased Grouse and other birds, are also eaten. The Eagle pounces on these creatures unawares, or even drives them over cliffs—never flies at and strikes them with the dash and daring so characteristic of the true Falcons. In fact, as we wrote long ago, after a careful study of the habits of the Golden Eagle, the bird is more like a Vulture than a Hawk, and we were going to say almost as harmless. The usual note of this Eagle is a yelping or barking cry.

The Golden Eagle is a very early breeder, and probably pairs for life, seeing that the same eyrie will be occupied or used in turn year after year. A site for this is usually selected on some noble crag or precipitous cliff, generally in a cleft or where the rocks overhang. Trees nowadays seem to be deserted, possibly because they offer a smaller amount of security. Occasionally a sea cliff is selected; and we have a vivid remembrance of an eyrie in such a situation on the west coast of Skye, more especially because through a rotten rope we nearly lost our life in an attempt to reach it. The nest of the Golden Eagle is a massive, well-made structure—a pile of sticks and branches and pieces of turf, lined with dry grass, moss, and tufts of green plants, generally *Luzula sylvatica*. The two or rarely three eggs are dirty white or very pale blue in ground colour, blotched and spotted with reddish brown and lavender grey. Usually in each clutch one egg is much more richly marked than the other. In the last two eggs of the Golden Eagle which we blew from Scotland this was very noticeable, one of them being almost spotless. Both parents assist in incubating them. This Eagle sits very lightly, flying away from the nest at once, and never, so far as our experience goes, showing any inclination to attack a human intruder. The eggs are often laid long before the snow is off the mountains, in March or early in April — a circumstance which is fortunate, for the "collector" is seldom so far afield as the Highlands until a more genial season.

The Golden Eagle cannot readily be confused with any other British bird except the White-tailed Eagle, whilst from this species it is readily distinguished by its feathered tarsi. The general colour of the plumage is dark brown, often with a purplish sheen, except the nape, which is pale brown, and the tail, which exhibits grey mottling. The female resembles the male in colour. Young birds are specially characterised by having the basal half of the tail white; and the feathers of the body, especially on the lower parts, have white bases. The total length of an adult Golden Eagle is about thirty-four inches.

THE WHITE-TAILED EAGLE

(*HALIÆETUS ALBICILLA*)

WE are glad to say that this magnificent species still retains a place in the British avifauna, and, although recently threatened with complete extermination, has slightly increased in numbers of late years, thanks to the efforts which on more than one estate have been made to protect it. The White-tailed Eagle was formerly much more widely dispersed over Britain than is now the case; still, we trust that for years yet to come it may remain an ornament to some of the wildest and most romantic scenery our isles can boast. We have ample evidence to show that within the past hundred years this Eagle actually bred on the Isle of Man, and in the English Lake District so recently as 1835! Among other English stations that once could boast the eyrie of this Eagle may be mentioned Lundy Island, the Isle of Wight, and possibly Cornwall. In the Lowlands of Scotland it was once even more plentiful than in England, extermination progressing northwards, and naturally becoming slower in wild and remote areas far from the haunts of men. Eyries were situated on Ailsa Craig, the Bass Rock, and in Galloway and Dumfriesshire. In Ireland this bird was formerly widely dispersed, but trap, gun, and poison (to say nothing of the rascally collector) have done their work only too well, and its numbers have been greatly reduced. Scattered eyries exist at the present time in some of the wilder western districts. In Scotland the chief stronghold of the White-tailed Eagle is amongst the Hebrides—in Skye particularly, also in Eigg, Scalpa, North Uist, Benbecula, the Shiant Islands, Rum, and Canna. Formerly the bird bred on St. Kilda; but the natives of those lonely isles will not tolerate such a formidable-looking species, and it is ruthlessly destroyed. Farther north it breeds on the Orkneys and Shetlands, but we very much doubt if a single eyrie is now inhabited anywhere on the mainland of Scotland. As this Eagle is a resident in the British Islands, the individuals of the species now dwelling in them are all that we are ever likely to receive, and it behoves us to see that the remnant of the indigenous stock is strictly preserved. This

Eagle is comparatively harmless, there is no excuse for its slaughter, and we hope that the senseless practice of laying poisoned meat, so commonly indulged in by shepherds to this bird's detriment, may soon be made an illegal one. There are at present enough Eagles left to restock many a now deserted district; and although we can never hope to see the big bird in any southern haunt again, we may do our best to protect it where preservation is possible. The White-tailed Eagle has a very wide range beyond our limits, being found from Greenland to Kamtschatka. It breeds in many parts of Northern and Central Europe, from Scandinavia to the valley of the Danube and Turkey; whilst in winter it visits North Africa, from the Canaries to Egypt, in which latter country it is said also to breed. Its winter quarters in Asia include India, China, and Japan.

In our islands now the favourite haunts of the White-tailed Eagle are maritime ones, but in other countries the bird appears to have as much preference for inland localities. No other scenery in Britain excels in grandeur that of this Eagle's haunt—the wild mountainous islands of the north, with their secluded lochs and long ranges of sea-washed crags, their bare hills and stream-pierced dales and glens, all offering that solitude and freedom from molestation in which the bird delights. There is a considerable amount of similarity between the habits of this and the preceding species. Both birds are sluggish, heavy, and we might almost say ungainly upon the ground, but in the air they become majestic. The flight of the White-tailed Eagle is marked by the same characteristics as that of its ally—the same high soaring in immense circles, the same gliding motions relieved by occasional flaps of the mighty pinions, the same descents from the clouds on uplifted wings. It is a solitary species, save in the breeding season, and wanders far and wide over large stretches of country in its quest for food. This consists largely of carrion and diseased and weakly animals and birds, such as lambs, hares, Ducks, Ptarmigan, and sea-fowl. The bird also feeds on fish, which it either catches for itself or finds dead and stranded along the shore. Its note is a yelping or barking cry.

The White-tailed Eagle breeds equally as early as the Golden Eagle, and its eggs are laid in March or early in April. So far as our islands are concerned, this species appears now always to select a maritime cliff for nesting purposes, and some of these that we have had the pleasure of visiting are magnificent to a degree. So far as our experience goes, the bird always selects a site for its eyrie which commands a wide range of country. Some of these nests are built in rocks absolutely inaccessible to man; others in situations which can be reached by even a moderate climber. The nest is a huge mass of sticks, often the accumulation of years, generally lined with dry grass, bunches of wool, and leaves of some green plant. Some nests are much more elaborate than others, the site influencing this to a great extent. We have seen nests which occupied an entire crevice in the cliffs, the hollow being filled up in the same way that a Jackdaw will do; whilst others on the flat ledges were composed of not more than a quarter of the material. In some countries, we might say, the nest is made in a tree or even on the ground. The two eggs are white, and generally without markings. Incubation is performed by both sexes, and but one brood is reared in the year. An inexperienced person might think that to approach the nest of such a big bird would be a somewhat risky undertaking; but the sitting Eagle flies away almost as soon as it is disturbed, and appears to show no further interest in the unwelcome visit. A Ring Ouzel is immeasurably more plucky and pugnacious at its threatened nest. The impression that Eagles are fierce and courageous is a widely prevailing one, yet a very erroneous one; and in this respect they more closely resemble the Vultures than the Falcons and Hawks. We should also state that this Eagle most probably pairs for life.

The general colour of the upper parts of the White-tailed Eagle is brown, paler on the head and nape, which in very old birds are almost white; the quills are nearly black; the under parts are very dark brown; the tail is pure white. The female resembles the male in colour, but is slightly larger and darker. The young bird is much darker than the adult, and the plumage is more mottled; the tail is dark brown. The total length of the male of this Eagle is about twenty-nine inches, the female four or five inches more.

THE HONEY BUZZARD

(*PERNIS APIVORUS*)

THE reason why we have not included the Honey Buzzard in our account of extinct species is because we believe that the bird still breeds within our area. We fear, however, that there is little hope for saving the bird from extinction. Its fate rests solely on the protection of the one or two pairs that still visit us in spring; when these are gone, the Honey Buzzard will become as extinct in England as the Crane and the Spoonbill, and with as little chance of its being restored. Of all our indigenous birds of prey, the Honey Buzzard seems to have suffered the most from the various exterminating forces which have been operating during the past century or so. There can be doubt that this bird formerly bred in many parts of the British Islands. Willughby tells us that in his day it was by no means uncommon. The last stronghold of the Honey Buzzard appears to be the

New Forest, and here still, we believe, a few pairs linger, in spite of the heavy price that has been set upon their heads by "collectors." The poor bird is one of the most harmless of our native species, and its good offices in destroying wasps should claim for it immunity from persecution, to say nothing of its beauty and the charm it lends to woodland scenery. We can attribute the extermination of the Honey Buzzard to nothing but the persecution of ignorant gamekeepers and the stupid craze for British-taken eggs and skins. Of all our threatened species none stand in greater need of protection, and whatever steps may be taken to save it must be prompt and effective.

Beyond the British Islands the Honey Buzzard is somewhat sparingly and locally distributed as a breeding species over the greater part of Europe, from the Arctic Circle southwards to the Pyrenees and Bulgaria. Eastwards its breeding range extends from Asia Minor and Turkestan, across Southern Siberia and North China to Japan. It passes the Mediterranean countries on migration only, and its winter home includes the African portion of the intertropical realm, and possibly India and Siam.

The Honey Buzzard passes into Europe, often in large flocks, about the middle of April, the migration lasting for about a month, and returns south during September and October. The British individuals arrive in our area early in May. Its habits are somewhat similar to those of the Kite and the Common Buzzard. It is a somewhat sluggish bird, spending much of its time upon the ground, where it is said to run with remarkable speed and grace. When in the air, however, it often indulges in those soaring flights and sailing motions so characteristic of the raptores. Its note is a Buzzard-like cry, an oft-repeated querulous sound, seldom heard, however, except in the breeding season. The food of this species largely consists of wasp grubs, to obtain which it will dig into the ground with great perseverance, apparently utterly oblivious of the angry insects. Grasshoppers, frogs, lizards, mice, worms, and small birds are also eaten.

The breeding season of this bird begins early in June. Like the Kestrel, it does not make a nest for itself, but selects the deserted home of a Crow, a Magpie, a Kite, or a Common Buzzard, in which to lay its eggs; but before doing so it appears to re-line the structure with a quantity of green leaves, or twigs with the leaves upon them. This lining seems to be renewed from time to time as incubation proceeds. The eggs are usually two in number, but three and even four have been recorded. They are very beautiful objects, almost round, cream or pale red in ground colour, blotched and spotted with rich brown, often so thickly as to hide all trace of the paler ground. But one brood is reared in the season, and both parents assist in the task of incubation.

We are not aware that the Honey Buzzard is at all gregarious during the breeding season, although the bird migrates towards its nesting-grounds in flocks, and returns in the same way —a habit indulged in by several other raptorial species.

The Honey Buzzard may be readily distinguished by its densely feathered lores and its finely reticulated tarsi. The adult male has the head ash grey, the remainder of the upper parts brown; the under parts are nearly uniform white, with a few brownish bars on the chest and flanks; the tail is pale brown, crossed with three nearly black bars. A melanistic form of this bird is known with the under parts dark brown. The female resembles the male in colour, but wants the grey head. Young birds have pale margins to the feathers of the upper parts, and the under parts are streaked instead of barred with brown. The total length of the Honey Buzzard is about twenty-four inches.

THE MARSH HARRIER

(*CIRCUS ÆRUGINOSUS*)

HERE again we have the sad record of a species, once fairly dispersed over the British Islands, now confined to one or two localities, where it manages to elude that senseless persecution which seems likely to reduce it to extinction. We fear there can be little doubt that the Marsh Harrier breeds but in one English county at the present time, and not at all in Scotland. In Ireland the bird is very probably more abundant than it is in England, the country being less populated and far more suited to its requirements. We have evidence to show that this Harrier formerly bred in Devonshire, in Somerset, Dorset, Shropshire, Lancashire, and Yorkshire, possibly also throughout the marshy wastes of East Anglia. Whether this species ever bred in Scotland seems by no means clear. The only place in which the Marsh Harrier is now known with certainty to breed is in the Norfolk Broads. In Ireland, Mr. Ussher informs us that it still breeds sparingly in Queen's County and Galway, and "probably" in King's County and Westmeath. There can be little doubt that the drainage and enclosure of marshy lands and fens has had a great deal to do with the extermination of this Harrier in England; as poison and gamekeepers are chiefly responsible for its present rarity in Ireland. If the bird's haunts are destroyed, the birds perforce must go too; and possibly the day is not far distant when the Irish bogs will be the sole retreat of the Marsh Harrier in Britain. There, however, some means should be taken to ensure the bird greater security than it now enjoys.

Outside our area the Marsh Harrier has a very wide distribution, reaching across Europe and Asia to Japan. It is not an Arctic bird, breeding in the south of Sweden only, but it is pretty generally distributed over Temperate and Southern Europe, as well as throughout the Mediterranean countries of North Africa (in winter reaching to the Equator). It is a summer migrant in the northern areas, but a resident in warmer and more southern localities. Eastwards we trace

THE MARSH HARRIER

it from Asia Minor across Turkestan and Siberia to North China and the Japanese Islands in summer, and in winter southwards to India and South China. The presence of an allied form in Asia renders the definition of its limits in this direction extremely difficult.

The Marsh Harrier well deserves its name. It is a dweller in the wilderness of swamps, fens, wet moors, and marshy lands, and the inundated banks of slow-running rivers and weed-choked meres. In common with its congeners, it possesses the habit of beating to and fro in slow and somewhat laboured flight over these swampy wastes in quest of food, seldom pursuing its quarry like a Hawk, but dropping down upon it unawares. It is fond also of sitting on walls, big stones, or even trees, whence it frequently sallies to capture prey. It spends a great part of its time in the air, possessing enduring wing-power, which, however, is rarely exerted beyond a slow and measured flapping, the bird all the time intent on scanning every inch of the ground below. We have watched it thus for a long time passing up and down over a comparatively small extent of marsh in eager quest. Its habit of attending sportsmen and carrying off dead or wounded birds under the very noses of the dogs has often been remarked. The food of this Harrier is largely composed of small mammals (such as mice, moles, young rabbits), frogs, and small or weakly birds. It is also a great robber of eggs and nestlings, most adept at finding nests and spoiling them of their contents. It is said also to eat fish. The note of the female of this Harrier is described by Naumann as a clear *pitz* and *peep*: that of the male as *koi* or *kai*. The Marsh Harriers that breed in England and Ireland are probably residents.

The Marsh Harrier, for a raptorial bird, is a somewhat late breeder, its eggs not being laid before May in this country, but in more southern haunts in March. The bird is far too rare in the British Islands to breed gregariously; but in Spain, where it is very abundant, Irby records as many as twenty nests within three hundred yards of each other. Montagu states that he has found the nest of this Harrier in a tree, but the usual place is upon the ground amongst the reeds or even in shallow water; and it is said a deserted nest of a Coot or Moorhen is sometimes utilised. Seebohm records a large nest absolutely floating amongst the reeds in water several feet in depth.

THE MARSH HARRIER

As is usual with birds nesting in such aquatic situations, the structure is added to from time to time, not only to increase its bulk and stability, but to replace material that may be washed away. The nest of this Harrier is a bulky one, made of reeds, sticks, and twigs, and lined with dead grass and other aquatic vegetation. The eggs are from three to six in number, and pale bluish green, almost white in colour, occasionally marked with rusty brown. Incubation appears to be performed by the female, and but one brood is reared in the season.

The adult male Marsh Harrier has the head and nape creamy white, streaked with dark brown; the remainder of the upper parts is reddish brown, with paler margins to the feathers; the primaries are black, the secondaries and tail ash grey; the under parts are chestnut brown. The female, although a trifle larger, closely resembles the male in colour. Young birds are uniform dark brown, spotted with paler brown, except the crown and throat, which are pale buff. The length of this Harrier is about twenty-two inches.

MONTAGU'S HARRIER

(*CIRCUS CINERACEUS*)

ALTHOUGH Montagu can scarcely be credited with the honour of discovering the Harrier which now bears his name, for the bird was unquestionably known to and described by Linnæus, there can be no doubt that he was the first naturalist to show that the bird was a British species, and to clear up much confusion which then existed concerning another Harrier also breeding within our area. The evidence concerning the past distribution of Montagu's Harrier in the British Islands seems to show that the bird was never more than a fairly common summer visitor to the southern and eastern counties of England, and a rarer one to Wales, and as far north as the Solway district, in the south of Scotland. A hundred years ago this species was very much more common than it is now, although comparatively recent instances of its breeding are known in Devonshire, Somerset, Dorset, and Hants. Its principal haunts at the present day appear to be the heaths of Norfolk. Possibly the bringing of common land into cultivation may have had some influence in reducing the numbers of this Harrier; but there can be no doubt that the persecution of gamekeepers has had infinitely more. If we are to retain this elegant and pretty bird in our fauna, measures will have speedily to be taken, for all the available evidence at the present day goes to show that this Harrier is upon the very verge of extinction. The old stock of birds that has been in the habit of migrating to Britain to breed is just upon exhausted, and if the few remaining pairs are not shown some consideration, the species must cease to exist as a British one. This Harrier never seems to have been a regular inhabitant of Ireland, and only one or two odd birds have been obtained there.

Outside the British Islands Montagu's Harrier is generally distributed as a breeding species over Continental Europe, south of the Baltic and the Gulf of Finland. Eastwards we trace it as a breeding species into Turkestan and Southern Siberia at least as far as the valley of the Yenisei. The winter range of this species not only includes various parts of South Europe, but Africa (where a few are said also to breed in the extreme north) south to the intertropical realm, and in Asia, India, Ceylon, and Burma.

Montagu's Harrier is often seen in large flocks on migration; and wherever the bird has not been persecuted, more or less social tendencies are shown throughout the year. These birds pass into Europe from the south very early in spring, the individuals breeding in the British area reaching us in April. The return journey takes place in September and October. The favourite haunts of this Harrier are extensive heaths and commons, grain lands, and marshes. Its habits are very similar to those of the preceding species. The bird has the same easy, graceful flight, the same peculiarity of systematically hunting the ground by passing to and fro and dropping down upon its prey. At times it will glide for a long distance with outspread motionless wings, or hover for a few moments in a fluttering manner like the better-known Kestrel. This bird appears to spend a good deal of its time upon the ground. Like its congeners, it is a persevering searcher after the nests of other birds, and robs them of their eggs and young, or even pounces down upon the sitting birds. Its other food includes mice, moles, frogs, grasshoppers, locusts, snakes, and lizards—a bill of fare that bears eloquent testimony to the bird's usefulness to man. The breeding season of Montagu's Harrier is in May. The nest is invariably made upon the ground, and as certain localities are used annually the probability is the birds pair for life. A bare spot amongst the heath or furze is usually selected, and here a slight nest of straws or dry grass surrounded with a few twigs is formed, the whole usually being arranged in some slight hollow. The eggs vary from four to six, and are pale bluish white, occasionally with a few rusty brown markings. These eggs are apparently laid at intervals of a few days, but the bird begins to sit as soon as the first is laid. Incubation seems to be performed by the female alone. Saunders relates that a female flushed from her nest in the Isle of Wight flew away in ever widening circles, and returned in a similar circuitous manner, until close to her home she dropped upon it with closed wings. During the breeding season this Harrier may often be seen playing and toying with its mate high in air.

The adult male Montagu's Harrier has the general colour of the plumage grey, with black primaries and a black bar across the secondaries ; the outer tail feathers are marked with reddish brown and white bars. The under parts below the breast are white, streaked with reddish brown. The female is nearly uniform brown on the upper parts, streaked with rufous on the head and neck ; the wings and tail are brown, some of the quills in the former and all the latter barred with darker brown; the under parts are white streaked with rufous. The young somewhat closely resemble the female, but the upper plumage shows more buff margination, and the under surface is buff streaked with reddish brown. The total length of this Harrier is about eighteen or nineteen inches. Montagu's Harrier may be distinguished in any plumage by the absence of a notch in the outer web of the fifth primary.

THE HEN HARRIER

(*CIRCUS CYANEUS*)

THE trivial name of this Harrier is a significant testimony to its former abundance in the British Islands. Even at the present time we should class it as the most common of the three British species, notwithstanding a long course of persecution, and very probably because its haunts are inaccessible to the multitude. At one time very widely dispersed, it now seems to be confined to the wild moorland districts from Cornwall and Devonshire through Wales to the Lake District, and thence northwards to the Highlands, the Western Isles, the Orkneys, and the Shetlands. There is evidence to show that the Hen Harrier was formerly a dweller in the fens of East Anglia, but has now become extinct there, as it also has in many moorland districts of the west and north. In Ireland it is still found as a breeding species, though in sadly reduced numbers; and, notwithstanding the suitability of the country for this species, it is said by Mr. Ussher only to nest, and that sparingly, in Kerry and Galway, possibly in Antrim, Queen's County, Waterford, and Tipperary. Drainage and enclosure of waste lands has probably contributed to the decrease of the Hen Harrier in lowland and cultivated areas, but its disappearance cannot be attributed to such a cause in the moorland and mountain haunts it was known to frequent. Here, as indeed elsewhere, it has been ruthlessly shot down and trapped by gamekeepers and landowners. It is more than probable that the Hen Harrier will soon be banished entirely to the mountainous moors.

The extra British breeding range of the Hen Harrier extends across Europe and Asia to the north island of Japan, from about the limits of forest growth in the north down to Spain, Central France, the Alps, Carpathians, Turkey, South Russia, Palestine, and Southern Turkestan in the south. Its winter range includes the basin of the Mediterranean, Northern India, Mongolia, China, and Southern Japan.

There can be little doubt that the Hen Harrier in its prosperous days was principally a summer visitor to the British Islands, although a few birds appear to remain over the winter in some districts. From France northwards on the Continent the bird is a regular migrant, moving to its breeding-grounds in March or April, and returning in September, October, and November. Its favourite haunts in Britain are wild moors and heaths and the rough scrub-covered sides of mountains. During migration it is to some extent gregarious, but at other times appears to live solitary or in scattered pairs. Its habits very closely resemble those of the preceding species. It shows the same peculiarity of flying slowly up and down its haunts, at no great height, searching the ground below for the objects on which it subsists. It is also very regular in its movements, searching particular places about the same hour each day, and passing over the country by certain routes. It is a great robber of nests, especially those of the smaller birds, feeding upon the eggs and nestlings, and even the parents, when able to drop down upon them unawares. Unlike the other species, it is said often to chase its quarry on the wing. Its other food consists of small mammals, such as mice and moles, of frogs and lizards, grasshoppers, locusts, and other insects, whilst its partiality for chickens has long brought it into ill-repute with the poultry-keeper. The nest of the Hen Harrier is always made upon the ground, often amongst long heather or gorse, less frequently on barer ground. A nest of this Harrier we examined in Skye was made in an almost impenetrable heather thicket; and we were assured by gamekeepers in the island that sheep broke many eggs of this bird in wandering over the hills. The nest is usually a mere hollow scantily lined with dry grass and surrounded with a few twigs. Sometimes the nest is much larger, a foot or more in height, yet made of similar material. The eggs are from four to six in number, bluish white, rarely marked with a few rusty spots. The Hen Harrier is a very conspicuous bird on the mountain moors, looking like a Gull in the distance, and its slow, measured flight increases the similarity. But one breed is reared in the season, and the eggs are incubated by the female. The note of this species has been described as an oft-repeated *ker*.

The adult male Hen Harrier has the general colour of the upper parts and the breast a clear slate grey: the rump and the under parts below the breast are white; the quills are black, but the tail is grey, like the upper plumage generally. The female has the general colour of the plumage brown, palest on the under parts, which are streaked with reddish brown; the upper tail coverts are white as in the male, but marked indistinctly with brown; the tail is dark brown, barred with buffish brown, and tipped with pale buff. The young somewhat closely resemble the female in colour. The total length of this Harrier is about nineteen or twenty inches, females being a trifle larger than males.

THE DOTTEREL

(*EUDROMIAS MORINELLUS*)

WHETHER the Dotterel ever bred on any of the hills in the south of England does not appear to be certainly known; but there is abundant evidence to show that in former times the bird occurred in some abundance during the season of its migrations on the hills and downs bordering the English Channel and elsewhere farther inland. Possibly some of these remained to breed. Nowadays the Dotterel is not only rare on passage, but has been well-nigh if not completely exterminated in many of its British nesting-places. In this case we cannot exactly lay the blame of extermination to the collector; although we have reason for stating that its eggs are sought eagerly by oologists and dealers, especially now the bird has become "rare" and British-taken eggs are at a premium. The Dotterel is now so rare because it has been slaughtered so wantonly, not only for its flesh, which is or was considered a great delicacy, but for the sake of its feathers, which are used in the making of artificial flies for fishermen. As the bird was extremely fat, especially in spring, it was caught before it had time to breed, and hence its numbers gradually diminished. The bird still breeds, we believe, on the hills in the Lake District as well as on the Cheviots, but in numbers that are decreasing. Farther north, we are glad to say, it breeds in greater numbers on the hills of Dumfriesshire, on the Grampians in North Perthshire, and on the borders of Inverness-shire, and in Ross and Banff-shire. It has been found nesting in the Orkneys, but appears only to pass the Shetlands on migration. Elsewhere in our islands the Dotterel can only be regarded as a casual visitor or a passing migrant. If the Wild Birds Protection Acts were better enforced, there can be little doubt that the Dotterel would increase in numbers in Britain; but otherwise the species is bound to decrease and possibly become extinct. Beyond our limits the Dotterel has a wide range, breeding on the tundras above the limits of forest growth across Europe and Asia, and at high elevations on the Alps. In winter it is found in the basin of the Mediterranean,

—but sparingly on the European side,—and is an occasional visitor to the Canaries.

The Dotterel is a late migrant, not reaching its British haunts before the end of April or early in May, whilst the return passage extends normally over September and October. This species is not a dweller on the coasts, but loves the inland pastures, sheep walks, and bare mountain slopes. It is certainly a social bird, not only on migration but at its breeding haunts, and becomes even more so after the young are reared. All observers agree as to its exceptional tameness, this trustfulness having gained for it the name of "foolish" Dotterel. Its flight is rapid and powerful; but the bird spends most of its time upon the ground, running about in quest of food. This consists of insects, worms, grubs, snails, and the buds and shoots of plants. Its call-note is a plaintive *düt* variously modulated, and in the pairing season is prolonged into a kind of trill. In the British Islands the eggs of the Dotterel are laid towards the end of May or early in June. The nest is a mere hollow in the moss or grass on the uplands. The eggs are three in number, varying from yellowish olive to pale buff, richly blotched and spotted with dark brown, paler brown, and grey. The male—in this species as in the Red-necked Phalarope—is not so fine or showily dressed a bird as the female, and consequently the greater part of the duties of incubation and tending the brood devolve upon him. But one brood is reared in the year. At the nest the old birds frequently try by cunning artifices to lure an intruder away.

The adult female Dotterel in breeding plumage has the general colour of the upper parts greyish brown, becoming brownish black on the crown; the shaft of the first primary is white; the wing coverts and innermost secondaries and scapulars are margined with chestnut; the outer tail feathers have broad white tips. From the base of the bill extending backwards round the crown is a white stripe; the chin and throat are white; the breast is greyish brown; across the chest is a white band margined with black; the remainder of the under parts are chestnut, shading into nearly black on the belly, and buffish white on the thighs, vent, and under tail coverts. The male is smaller and less brilliant than the female. The young bird has the crown dark brown with pale margins, the breast mottled with greyish brown, the white gorget only faintly indicated, and the rest of the under parts white. The total length of the female

Dotterel is about nine inches. It should be stated that there is some difference of opinion respecting the colour and size of the sexes in this species. The matter seems to require further investigation.

THE KENTISH SAND PLOVER

(*ÆGIALOPHILUS CANTIANUS*)

THIS pretty species appears to have been at all times an excessively local one. It must always have an exceptional interest for British naturalists, inasmuch as it was first made known to science from examples obtained on the south coast of England little more than a hundred years ago. To Mr. Boys of Sandwich belongs the credit of its discovery. This gentleman sent an example to Latham, which was figured by Lewin in his work on British Birds published in 1800; whilst a year later Latham himself described it in the Supplement to his celebrated *Index Ornithologicus*, having received two more examples from Mr. Boys in 1791. Although this Plover has been obtained accidentally in other parts of the British area, its normal distribution is confined to the shingly beaches of Kent and Sussex. There is no evidence of its breeding on any other part of our coast-line, although the bird is fairly common in the Channel Islands. The present rarity of the Kentish Plover is entirely due to the greed of collectors, and it seems to us a monstrous thing that such is the case. If some means are not quickly devised for its protection, nothing can save the Kentish Plover from absolute extinction in the British Islands. The bird only requires protection during the breeding season, from April onwards, and we would make it illegal to shoot Kentish Plovers until the beginning of October, instead of the first of August, by which date the poor harassed birds would have retired south to their winter centres. The taking of the eggs should also be made illegal. No species more urgently needs protection.

The Kentish Plover is a summer migrant to the beaches of Western Europe, from France northwards to the south of Sweden. It is a resident on the coasts of the Iberian Peninsula, the Azores, Canaries, and Madeira, and along both sides of the Mediterranean. Eastwards, we find it frequenting the marshes on the South Russian Steppes, the beaches of the Black, Caspian, and Aral Seas, and those of the salt lakes in Turkestan, Dauria, and Mongolia. The winter range includes the coasts of

Africa south to the intertropical realm, the Mekran coast, the Indian Peninsula, Burma, the Malay countries, China, and Japan. The presence of several allied forms in Asia makes the definition of the winter area of this species somewhat difficult.

The Kentish Plover is seldom found far from salt water, either on the rough sand and pebble-strewn beaches of the sea, or on similar ground by the margins of salt lakes farther inland. This Plover arrives on the British coasts towards the end of April or early in May. Its favourite resorts are sandy beaches interspersed with patches of shingle and pebbles. Here its actions are very similar to those of the better-known Ringed Plover. It searches for food on the very margin of the incoming tide, running daintily hither and thither, or standing for a moment quite still, until the next spent wave causes it to trip lightly out of the way. The poor little bird is too rare in England now to display many social tendencies during the summer, the few scattered pairs keeping to their own particular haunts; but in autumn parties may sometimes be seen, broods and their parents migrating together. The flight of this species is very similar to that of the commoner Ringed Plover, rapid and well-sustained, and often accompanied by a series of shrill, oft-repeated notes. The alarm-note may be expressed by the syllable *ptirr*; the more usual call-note is a loud, clear *whit*. This latter note, during the pairing season, is often repeated so rapidly as to become a trill, and is uttered as the cock bird soars and flies in circles above his mate upon the sands below. The food of this Plover consists of crustaceans, sand worms, molluscs, and insects.

The Kentish Plover probably pairs for life, and returns season after season with admirable persistency to the same strip of shingle to breed. The eggs are laid towards the end of May or early in June. Nest there is none beyond a little hollow in the sand or shingle, whilst sometimes the eggs are laid on a drifted heap of dry seaweed. These eggs are usually three, but sometimes four in number, and are buff in ground colour, blotched, streaked, and spotted with blackish brown and grey. Few birds sit more alertly, and the moment danger is detected the wily parent runs from her charge for some distance ere rising. The young birds are very nimble, and when alarmed hide themselves by crouching low amongst the pebbles. Dr. Sharpe, who has had an enviable experience of this rare bird, thus writes respecting the young: "I have, however, captured several nestlings by resting my head on the shingle, when the little creatures become distinctly visible against the sky-line, as they run along with wonderful swiftness for such tiny objects. I could never bring myself to kill any of these fluffy little balls of down, with their great dark eyes and abnormally long legs; and later in the autumn I have been rewarded by seeing flocks of Kentish Sand Plovers feeding on the green herbage which skirts the harbours after the tide has receded. I once saw, from behind my shelter of a mud-bank, more than forty of these pretty birds feeding on the green moss near Romney Hoy, and a more interesting sight can scarcely be imagined." As will be seen from the foregoing particulars, the Kentish Plover becomes gregarious in autumn, as so many other kindred species do. This Plover rears but one brood in the summer, and the migration south begins in August and continues into September.

The adult male Kentish Sand Plover has the forehead and eyebrow white; the lores and a broad streak behind the eye black; another black patch separates the white on the forehead from the buff of the top of the head and the nape; the remainder of the upper parts, including the six central tail feathers, are greyish brown; the quills are dark brown, with white shafts to the primaries and concealed white bases to the innermost; the innermost secondaries are also margined with white; the remaining tail feathers are white. The general colour of the under parts is white, except a black patch on each side of the chest. The female resembles the male in colour, but the black on the fore crown is wanting; the breast patches are brown, and the buff on the head is not so extensive or rich in tint. In winter the buff is entirely wanting from both sexes; young birds resemble adults in winter plumage, but the dark feathers have pale margins. The total length of this Plover is between six and seven inches. It may be distinguished at all ages not only by its white nuchal collar, but by its interrupted pectoral band and black legs.

THE RUFF

(*MACHETES PUGNAX*)

IF this curious species still manages to retain a place as an indigenous British bird, that is all that can be said for it. We are still loth to regard the Ruff as extinct in our islands as a breeding species, for possibly it may yet be saved to us if the law already in existence for its protection be strictly enforced. The Ruff was formerly a very common summer visitor to the marshes of East Anglia, but is only known now to resort to a few localities in Lincolnshire and Norfolk. Professor Newton says there is but one locality left. Mr. Saunders states that a hen bird was shot from the nest as recently as 1882 in the former county, and also that a few pairs succeed in rearing their broods in the latter county. As we have found to be the case with several other species, numbers of Ruffs pass our islands on migration, but even these are dwindling in amount.

These passing migrants, however, are of no service in recruiting the indigenous stock, and as soon as that becomes extirpated, the Ruff as a breeding species will be lost to us for ever. Formerly the Ruff was so plentiful in the Fens that it was regularly snared and fattened for the table; but the drainage of these vast areas has robbed the bird of its home for the most part, and senseless, wanton persecution is doing the rest. In many respects the Ruff is one of the most singular of known birds, and one deserving of every effort being made for its retention in the British avifauna. There are many tracts of land still left suited to the bird's requirements; all that is necessary is to protect it, especially during the breeding season.

Beyond our limits the Ruff is a very wide-ranging species, being found during the breeding season over the greater part of Europe and Asia. In Europe it is said to breed as far north as land extends, and as far south as the valley of the Danube; in Asia, up to similar limits, across the continent to Kamtschatka, and south to the Kirghiz Steppes, Western Dauria, and possibly the valley of the Amoor. It is a well-known migrant in the basins of the Mediterranean, Black, Caspian, and Aral Seas, and winters in the African portion of the intertropical realm, in Northern India and Burma. Abnormal migrants of this species have been known to wander to South America, Borneo, Canada, and elsewhere.

The Ruff begins its migrations into Europe as early as January, and continues them until near the end of May. The return passage takes place in August, September, and October; but a few odd birds are often known to pass the winter on the British coasts. The Ruff is gregarious, not only on passage and in winter, but practically throughout the breeding season. During the non-breeding season the Ruff frequents mudflats and salt marshes on the coast as well as inland districts, but in summer its favourite resorts are swampy moors and rough wet ground, clothed with a carpet of coarse grass, hummocks of sedge, and rushes. The flight and general actions of the Ruff are very similar to those of wading birds in general. Its food consists of insects and larvæ, worms, snails, small seeds, and various vegetable fragments. Its note is described by some observers as a low *whit*, by others as *ka-ka-kuk*.

By far the most interesting portion of the Ruff's economy is that relating to its reproduction. The Ruff is polygamous, and, like most birds practising polygamy, the males are excessively pugnacious, and fond of displaying those curious nuptial plumes which render this species absolutely unique amongst Aves. During the mating season the males "hill," as it is termed—that is, resort to certain spots to engage in combat; and these battles are continued at intervals—generally in the morning—until the females retire to incubate the eggs. The males now take no further interest in the hens, leaving them to bring up the brood, whilst they wander about in flocks until the migration period arrives. Some very interesting particulars concerning the "hilling" of this species have been contributed to the *Ibis* by Mr. A. Chapman, who found the Ruff very common in the marshes of Jutland in the season of 1893. He writes: "It was with the greatest interest that we watched these singular birds, in congregations of from six or eight to twenty or thirty, beating their flanks with their wings, and otherwise performing the strangest antics. Often a pair of Ruffs would, with ruff and ear-tufts erect, stand facing each other for minutes together, their heads lowered and their bills nearly touching each other; then one would spring into the air and make a desperate rush at his retiring adversary, their aptitude for running over the ground at a marvellous speed being most extraordinary. Very frequently no Reeve was present during these exhibitions, and the persistency with which the birds refuse to be driven away from their selected 'hill' merits attention." After pairing, each female appears to select some spot for the nest away from her companions. This nest is made upon the ground in the swamps, and is generally placed in the centre of a tuft of sedge or coarse grass, which effectually conceals it. It is little more than a hollow in which a few dead leaves or bits of withered herbage are strewn. The eggs are four in number, varying from greenish grey to greyish green in ground colour, spotted and blotched with reddish brown and greyish brown. But one brood is reared in the year, the eggs for which are laid in May or early June.

The plumage of the adult male Ruff varies in colour to such an astonishing degree, that to attempt any detailed description in the space here available is absolutely impossible. We may, however, say that this variation is chiefly confined to the nuptial plumes which are assumed in spring—the ruff, the feathers on the breast and flanks, and the ground colour of the upper parts. An almost endless diversity or mixture of white, chestnut, and black with blue and green metallic reflections, is exhibited on these plumes, and it is interesting to remark that each Ruff assumes similar colours to those displayed in previous seasons. The wings are nearly uniform brown; the feathers of the lower back are brownish black, with chestnut margins; the under wing coverts and axillaries are white, as are also the centre of the belly and the under tail coverts; the tail is brown. The face in spring is bare of feathers, but covered with tubercles of various tints, said to correspond with that of the ruff or collar itself. The female —smaller than the male—wants all this decorative plumage, has the general colour of the upper parts black, each feather with a greyish-white or chestnut-buff margin; the feathers of the breast and flanks are brown, with pale margins; the remainder of the under surface is white; the wings and tail are similar to those of the male in colour. Young birds resemble the female, but the buff margins are more pronounced. Diagnostic characters of this species are the white axillaries, and the absence of white from the quills and central upper tail coverts. The length of the adult male is about twelve inches, the female two inches less.

THE RED-NECKED PHALAROPE

(*PHALAROPUS HYPERBOREUS*)

HERE again we have a most interesting and beautiful little species threatened with speedy extermination within the British Islands. Fortunately, its haunts are confined to the most remote areas, but even there the "trading collector" penetrates, and with results that may be readily imagined, seeing the price that British-taken birds and eggs command. There would be no thieves if there were no purchasers of stolen goods, and there would be none of these rascally speculative dealers ready to despoil the haunts and nests of our rarest birds, if egg collectors declined to purchase specimens which are literally costing the extermination of so many interesting birds. All the mainland haunts of the Red-necked Phalarope are now deserted. Formerly this species bred in many a Scottish shire,—in those of Perth, Inverness, and Sutherland for certain,—but nowadays its last remaining strongholds are on various islands on the west and north of Scotland, which it seems a pity more particularly to specialise. To watch these tame and gentle little creatures at their breeding stations on the wild islands of the north, is a sight whose charm no pen can do justice to: and it grieves us to think that continued persecution is rapidly bringing the day when such exquisite pictures of bird life will fade from our Scottish waters for ever. Even within the past ten years the number of breeding birds has sadly diminished, and there can be no doubt whatever that the indigenous stock is fast becoming exhausted.

Beyond the limits of the British Islands the Red-necked Phalarope has a very extensive range, breeding throughout the Arctic and sub-Arctic regions of both hemispheres. In America we find it from Alaska to Greenland; in the Old World from Iceland and the Faroes across Europe and Asia to Kamtschatka. In Continental Europe this Phalarope breeds as far south as the Dovrefjeld in latitude 62°, and in Eastern Asia as low as latitude 55° on the shores of the Okhotsk Sea. Its winter migrations extend in the Old World down to the basin of the Mediterranean, Persia, Northern

India, China, Malaysia, and Japan; whilst in the New World they reach Mexico and Central America.

The Red-necked Phalarope is quite as aquatic in its habits as a Coot, perhaps even more so, being seldom seen on the land for long together, except in the breeding season. It is an absurdly tame and confiding little bird, especially at the nest, and at all times seems more or less gregarious. This species swims well, with a buoyancy exceeded by no other bird. It is a pretty sight to watch its actions when swimming across some deep, clear pool, progressing in a more or less zigzag direction, each stroke of its feet accompanied by a nod of its head. It may also be seen running quickly and gracefully about the marshy shores, wading or swimming the intervening pools, or even tripping lightly over floating masses of weed. Its flight is not only rapid but powerful; and Seebohm remarked that when one was shot, its companions came and hovered above it, and then alighted near it, just as Terns will often do. The usual note of this Phalarope is a shrill, clear *weet*. Its food is composed chiefly of insects, but worms, crustaceans, and other small marine creatures are sought.

The Red-necked Phalarope reaches its breeding haunts in Scotland towards the end of April or early in May. Here its favourite nesting-places are on the banks of rush-fringed pools, which stud the moors at no great distance from the sea. As these places are visited year by year, it seems probable that the bird may pair for life. This Phalarope nests in scattered colonies, and throughout the breeding season may be seen in companies swimming on the water or standing or running about the marshy moors. The nest is slight, and either made upon the ground or a short distance above it in a tuft of coarse grass or rushes. It is little more than a hollow somewhat neatly lined with dry grass or scraps of sedge leaves and reed. The four pyriform eggs range from pale olive to buff in ground colour, blotched and spotted with umber brown, blackish brown, pale brown, and grey. But one brood is reared in the year, and the eggs are chiefly incubated by the male. It may be of interest to remark that in this, as in some other species, the female is larger and more showily attired than the male; she takes the initiative in courtship, and leaves her mate to take the greatest share in bringing up the brood. As possibly bearing on this curious fact, we may mention that

Messrs. Pearson and Bidwell, during their visit to Northern Norway, repeatedly saw one female attended by two males, and pertinently ask whether this species is polyandrous? The question is certainly worthy of further investigation.

The adult female Red-necked Phalarope in nuptial plumage has the head, the back of the neck, and the shoulders slate grey; the remainder of the upper parts of the body is grey; the wings are brown, the scapulars striped with chestnut, the innermost secondaries narrowly and the greater coverts broadly tipped with white; the tail is also brown, but the upper tail coverts are barred with white. The chin and throat are white, the front and sides of the neck chestnut, the upper breast grey, the remainder of the under surface white, flecked with grey on the flanks and under tail coverts. The male in nuptial plumage is much duller than the female, otherwise resembles her in colour. In winter plumage the chestnut and grey are absent from the neck, and the chestnut disappears from the scapulars; whilst all the grey feathers of the upper parts are margined with white; and the forehead and entire under parts are white. Young birds are brown on the breast, and the feathers on the forehead, mantle, scapulars, innermost secondaries, and upper tail coverts are dark brown, with chestnut margins. The tail is also brown, with similar edges. The total length of the female is about eight inches, the male about an inch less.

THE ROSEATE TERN

(*STERNA DOUGALLI*)

THE Roseate Tern is another species possessing more than ordinary interest to British naturalists, because it was first made known to science by Montagu, who described it and named it after its discoverer, from a skin which had been obtained by Dr. MacDougall of Glasgow on one of the Cumbrae Islands in the Firth of Clyde. The worthy Doctor found this species breeding sparingly in company with a large colony of Common Terns, and furnished Montagu not only with the specimen that he described in the Supplement to his famous *Ornithological Dictionary*, but with particulars of its habits and characteristics. This was between eighty and ninety years ago. Selby afterwards found it breeding on the Farne Islands, and it was also discovered breeding in various other parts of the British Islands. Its best-known resorts were these famous islets off the Northumbrian coast, but other stations were on the Scilly Isles, on Foulney and Walney off the Lancashire coast, as well as other islets off the coasts of Scotland and Ireland. Possibly the Farne Islands have never been absolutely deserted by the Roseate Tern, and though extinct now in most of its old retreats, it still breeds upon them, and is likely to continue doing so now that the birds upon them are being strictly preserved, purely, we believe, by private enterprise. In 1896, Dr. Sharpe was informed of another "nice little colony" established in Wales; so that reasonable hopes may be entertained of the beautiful Roseate Tern thoroughly re-establishing itself in our islands, after being apparently on the very brink of extinction. Great care, however, will be necessary, and the few resorts of this species kept as secret as possible, and free from the intrusion of trading and grabbing collectors. It is possible that the scarcity of this Tern is in a measure due to the persecution of man, but another cause, and a more serious one, may be found in the fact that the bird is driven off by the more powerful Common Tern. Mr. Saunders was assured by Dr. Bureau that no less than three colonies of the Roseate Tern had succumbed to the larger species on the coast of Brittany alone.

THE ROSEATE TERN

Beyond the British Islands the Roseate Tern has a very extensive range along the coast-line of the Atlantic and Indian Oceans. From the western coasts of France we trace it as a breeding species up the Mediterranean, in Tunis, and round the African coasts, thence to the Mascarene Islands, Ceylon, the Andaman Islands, the Malay Archipelago, North and West Australia, and New Caledonia. Returning to the Atlantic, we find this Tern recorded from the Azores, formerly breeding on the Bermudas, and nesting along the coasts and islands of Eastern America, from Central America, and the West Indies northwards to Massachusetts. The Roseate Tern is unquestionably a tropic species migrating north and south in the Old World to breed, but northwards only in the New World, so far as is at present known.

The habits of the Roseate Tern, so far as they have been observed, very closely resemble those of allied species. The bird is eminently a coast one, attached to the shore and the islands near it. To the British area it is a summer migrant only, and a late one, not reaching its breeding-places until nearly the end of May. Its flight and actions generally are very similar to the Common Tern; but its black bill and rosy-tinted under parts, its shorter wings and longer tail, render identification easy, and prevent any confusion with the better-known species. Its note is the usual *kree*. The food of this Tern apparently consists entirely of small fish, which it catches by dropping down upon them Gannet-like, or whilst supporting itself with rapidly beating wings just above the water.

The favourite breeding haunts of the Roseate Tern are low, rocky islands with sand and shingle beaches. No nest is apparently made in this country; but Brewer states that a little dry grass and seaweed are collected by the birds breeding in some American stations; whilst M. Blanc assured Mr. Whitaker that in Tunis, where this species has only recently been discovered nesting, grass bents occasionally line the hollow in which the egg is deposited. There is considerable diversity of opinion respecting the number of eggs laid by this Tern. Most authorities agree in saying two or three eggs form a clutch; but Mr. Proud (from his experience at the Welsh colony noticed above) asserts that never more than two are laid; whilst, lastly, M. Blanc in Tunis maintains that but one is laid. They vary in ground colour from creamy buff to buffish brown, blotched, spotted, and clouded with reddish brown and pale grey. As this Tern is a late migrant to arrive in Britain in spring, it is equally an early one to depart in autumn, flying south as soon as the young are able to fly.

The adult male Roseate Tern has the general colour of the upper parts pale slate grey, palest on the rump, upper tail coverts, and secondaries; the tail, which is deeply forked, is pale grey, the long slender outermost feathers nearly white; the crown and nape are black; the cheeks, throat, and entire under surface white, flushed with a delicate rose tint, which, however, fades sooner or later after death. The female resembles the male in colour. The young are barred with black on the upper parts; the head and nape are brownish black, streaked with white; the under parts want the rosy flush. This Tern may be recognised by the white inner webs of the primaries. The length is about fourteen inches.

THE GREAT SKUA

(*STERCORARIUS CATARRHACTES*)

THIS imposing species seems always to have been a particularly local one in the British Islands, and there appears to be no reliable evidence that it ever bred in any part of them except the Shetlands. One would have thought that in such a remote locality the bird would have been fairly safe, but of late years it has been mercilessly harassed by collectors, and at one time reduced almost to the verge of extinction. It is, however, most gratifying to record that stringent measures for its preservation were taken in time, with the result that it is now on the increase. All lovers of our British birds must feel grateful to the Edmonston family for their efforts to preserve and protect this at one time vanishing species, and rejoice in the success which has attended them. It is pleasing to know that such efforts have already been acknowledged and rewarded by the Zoological Society of London bestowing a silver medal on the Great Skua's preservers. The two colonies of this species are situated on Unst and Foula. In the spring of 1891, Mr. Thomas Edmonston engaged a special keeper to live for three months on Hermanness, " to keep watch and ward by night and day over the Skuas' home." Early in May nine pairs of Skuas returned to the ancient nesting-place, two pairs of which unfortunately settled beyond the sacred limits of protection, and their eggs in due course were stolen. The other seven pairs, thanks to careful and ceaseless watching, succeeded in rearing their broods. At the neighbouring colony of Foula about a hundred pairs of birds appeared in the spring of 1891, and although most of the eggs of the first laying were taken, about sixty young were reared out of the second attempt. Mr. Edmonston, we should say, is of the opinion that the Great Skua will not increase much beyond its present numbers, because the Lesser Black-backed Gull and the Herring Gull are decreasing, and on these species the Skua chiefly depends for its piratical livelihood. "Protection for the Skuas," he writes, "implies some measure of protection also for the Gulls; but unless the latter greatly increase, the former cannot be expected to do so. In existing conditions, and pending a possible large increase in the number of Gulls, it is nearly certain that the Skua colony can only be increased by enlarging the area of ground protected." Would that many another persecuted and fast-vanishing British species could find such protectors as the Great Skua has found in Shetland! To any other part of the British Islands the Great Skua is only a wanderer, and it is scarcely ever seen in Ireland at all.

The range of the Great Skua beyond the British area, although extending across the Atlantic, is comparatively a restricted one. The bird breeds in the Faroes and Iceland, but is said by Hagerup to be only occasionally seen in South Greenland; whilst in America it is said to breed near Hudson Strait. In winter it wanders down the West European coasts to Iberia and Morocco, but seldom passes through the Straits of Gibraltar; whilst on the American side it is said to wander as low as New England.

The Great Skua is a thoroughly oceanic species, gifted, like most of its order, with ample powers of wing. It may aptly be described as the feathered pirate of the northern seas, depending largely for food upon the Gulls, which it pursues unmercifully, and with great fierceness compels to drop or even to disgorge the fish they have caught. The Gulls dread the Skua almost as much as they fear the Peregrine; it follows them in their quest for food often for long distances from land, and by its greater powers of flight is able to chase and rob them at will. To a great extent this Skua is solitary in its habits, except during the breeding season, and even then it keeps much in pairs, although assembled in considerable numbers, as Saxby long ago remarked. The usual note of the Great Skua is an oft-repeated *ag, ag*; but under the excitement of chasing Gulls it utters a loud note, which has been likened to the word *skua* or *skui*—hence the bird's name. The food of this species consists largely of fish stolen from the Gulls; the bird will also catch them for itself. Wounded or weakly birds, especially the nestlings of other sea birds, offal from the fishing-boats, and even carrion on the beach, are also devoured.

By the end of April the Skuas that breed within our area begin to assemble at the old accustomed haunts, which are wild moorlands at no great distance from the sea. Numbers of nests are scattered over a comparatively small area, so that the bird must be considered a social one at this season. The nest seems always to be made upon the ground, and is generally little more than a hollow in the moss or turf, in which a few bits of dry grass have been arranged as a lining. The eggs are two in number, and vary from pale buff to dark buffish brown in ground colour, somewhat obscurely and sparingly marked with dark brown and greyish brown. All observers who have visited the breeding-grounds of this Skua have been impressed with the bold way in which it seeks to defend its eggs or helpless young. Fearlessly flying round the intruder's head, both male and female advance towards him, swooping down as if about to strike, and showing little fear even at the report of a gun. Dogs are beaten off the sacred spot, and even the powerful White-tailed Eagle or the Raven are glad to retire before such spirited and angry attacks. But one brood is reared in the season; but if the first clutch of eggs be taken, another will be produced. As soon as the young are reared, the breeding-places are more or less deserted, and for the remainder of the year the birds lead a maritime life, wandering far and wide over the surrounding seas in their piratical quest of food.

The general colour of the upper parts of the Great Skua is dark brown, mottled and streaked with paler brown, palest on the nape, which is clothed with somewhat pointed feathers; the quills are dark brown with white bases, very conspicuous when the wings are outspread; the tail also is brown, the feathers having concealed white bases. The under parts are pale rufous brown, streaked on the breast and flanks with darker brown. The female is said to be a little larger than the male, otherwise similar in colour. Young birds resemble their parents in colour, but are a trifle more marked with rufous on the back, and the feathers on the nape are not quite so pointed. The total length of this Skua is about twenty-two inches.

SOME THREATENED BRITISH SPECIES

WE may aptly bring the first part of the present volume to a close by a brief review of certain species which, though not exactly threatened with speedy extermination, are or have become sufficiently local to bring such a fate within the bounds of probability. In almost every case, the species concerning which these warning words are penned have most to fear from the persecution of man, from indiscriminate robbing of their nests, or slaughter of the old birds themselves for the sake of their skins. The professional dealer in objects of this description is greatly to blame, but we think the purchaser of his wares is worthy of greater censure.

Our first species is the Dartford Warbler (*Sylvia provincialis*), which is not only a very local bird, but one whose distribution in our area is extremely limited. It is a resident in most of the southern

SOME THREATENED BRITISH SPECIES

counties of England from Cornwall eastwards, thence northwards along the Thames valley and through some of the Midland districts—Worcestershire, Leicestershire, Derbyshire, to the extreme south of Yorkshire, where years ago we have taken its nest. A few may also breed in Cambridgeshire, Norfolk, and Suffolk. The Dartford Warbler is another of those species which, in the event of the indigenous stock becoming extinct, can never be replaced by normal means. This species has been considerably reduced of late years by severe winters — a contingency to which our summer migrants are not exposed. To this cause the late Henry Swaysland attributed its almost complete disappearance from the gorse coverts of Sussex. Collectors of birds and eggs also harass this interesting little Warbler not a little. Fortunately, it is of secretive habits, and its nest is very difficult to find; but, notwithstanding these facts, the bird should be carefully protected during the breeding season, and the taking of its eggs made illegal in the several counties which it frequents.

Our next threatened species is the Chough (*Pyrrhocorax graculus*). This species was formerly much commoner and more widely dispersed than it is now, and though "once upon a time" a dweller in inland localities, at the present day maritime cliffs are almost its sole remaining stronghold. It still breeds, if in diminishing numbers, from Dorset west to Cornwall. A few birds breed on Lundy Island; colonies here and there exist along the rockbound coasts of Wales, as well as in one or two inland localities in that country; a few still nest in the Isle of Man, and possibly in Cumberland. Up the west coast of Scotland it is fairly well established, especially on the Island of Islay, and in smaller numbers in Jura and Skye. In Ireland its chief resorts are along the coasts of Kerry, Mayo, Donegal, Antrim, Waterford, and Cork. It also still continues to breed on the Blaskets. The most singular thing about the decrease of this species is that it cannot fairly be attributed to persecution by man or the destruction of its ancient strongholds. Evidence is not wanting that the decrease of the Chough is contemporaneous with the increase of the Jackdaw in each particular locality, and it seems probable that the stronger Daw is ousting the Chough from its ancestral homes. We would suggest by way of experiment, that where these interlopers seem actually to be dispossessing the Choughs, a reduction of their numbers should be made. Collectors work some harm in the more accessible districts, whilst the Peregrine is credited with the work of extermination in others. The Chough is fairly well established in our islands at present; but the tendency towards decrease is certainly marked, and the species requires to be carefully watched by the preserving naturalist.

The Golden Oriole (*Oriolus galbula*), is the next species concerning which we have a few warning words to write. Ordinary readers are scarcely aware how frequently this handsome and conspicuous bird visits the British Islands, or that it has actually bred in them. So far as we can see, there is nothing to prevent the Golden Oriole becoming as common this side of the English Channel (as it most probably was in remoter ages) as it is on the other. The bird is said to be a regular spring visitor to the Scilly Islands and Cornwall, and thence onwards through the southern counties as far as Norfolk, but with perhaps lesser frequency. It must be remembered that such very showy birds have difficulty in penetrating far after once landing on such inhospitable shores as ours. Possibly this bird has bred in Kent, Surrey, Essex, Northamptonshire, and Norfolk. Mr. Harting records that a pair reared a brood at Dumpton Park in the Isle of Thanet in

1874, thanks to the protection and consideration shown them by the proprietor; and again returning the following year to meet with similar success. Possibly the poor birds were destroyed on migration before a third effort could be made. The fact, however, very clearly proves that there is a normal migration of this species to Britain, and every effort should be made to encourage and protect such handsome, musical, and interesting birds. Their beauty, alas! is a fatal attraction to every owner of a gun, to every "collector" of British birds; and until English people show more kindly forbearance, we are afraid the Golden Oriole's attempts to settle amongst us will be futile.

Next on our warning list comes the Hobby (*Falco subbuteo*), which, through being a summer visitor only to our English woodlands, is fortunately only exposed for half the time to that wanton persecution so persistently bestowed upon all our indigenous birds of prey. We have personal knowledge of the ruthless way in which the nests of this Falcon have been robbed over entire districts season after season, to supply certain dealers in birds' eggs, only too eager to meet the demand for British-taken specimens. To this wholesale taking of the eggs must be added the incessant persecution of gamekeepers, so that the only wonder is the Hobby exists as a British species at all. This pretty little Falcon arrives in those English woodlands where it breeds in small numbers in May. Its summer resorts in England are principally in the south-eastern and midland counties of England, including Hampshire, Essex, Suffolk, Norfolk, Cambridgeshire, Lincolnshire, Leicestershire, Northamptonshire, Derbyshire, and Yorkshire. It has only been known to nest on one occasion in Scotland. We may also add that the Hobby, for a raptorial bird, is a comparatively harmless one, its food consisting chiefly of insects and small birds; but the latter do not appear to be killed in any great numbers. Our stock of indigenous Hobbys may yet be far from exhausted; still, we have the fate of the Honey Buzzard and the Harriers before us, and it behoves us to afford the present species some protection before it is reduced to a mere remnant. We may here take the opportunity of alluding to the Goshawk (*Astur palumbarius*), and to state, in our opinion, that this species was never indigenous to the British area within historical time. Certainly there is no evidence for it which can be classed as thoroughly reliable, and there can be little doubt that this Hawk was never more than it is now, an abnormal visitor on migration.

Passing allusion might here also be made to Baillon's Crake (*Crex bailloni*), and the Spotted Crake (*Crex porzana*), the former of which may just possibly breed within our area; whilst the latter, although far less common than formerly, is still a regular summer visitor to various parts of the kingdom. Drainage and enclosure of swamps and fens has curtailed the haunts of these birds, and we express the hope that both species may be shown consideration by sportsmen and collectors.

We now have to appeal on behalf of that exquisite little bird, the Lesser Tern (*Sterna minuta*). It is a species that has sadly decreased in numbers during the last twenty years. From some localities it has entirely disappeared; from others it is rapidly vanishing. To a great extent the extermination of this species is due to the bird's habit of frequenting the coast rather than islands for nesting purposes. This places it absolutely at the mercy of every wandering rascal. Haunts of the Lesser Tern on the Lincolnshire coast that I knew years ago contained scores of pairs, are now deserted, and I attribute this to the rapid rise of certain wateringplaces in their vicinity. Season after season the poor little birds lost almost every egg, picked up by excursionists; year after year their once secluded shingles became the summer resort of crowds of despoiling "trippers," and the Lesser Tern has disappeared. This has gone on in many other places; but we are glad to hear that in some localities efficient steps are being taken to preserve this Tern from extinction. We are afraid this will be an exceptionally difficult task, owing to the habits of the bird: still, it should not prove an insuperable one. It is useless, perhaps, to appeal to seaside visitors, and we fear that in all the more populous parts of the coast where the Lesser Tern breeds, the bird sooner or later will become extinct. We might add that a great many Lesser Terns have been shot for the sake of their plumes, the bird from its small size being in great requisition by milliners.

A few passing words must now be said for the Divers. We have at least two species of these breeding within our limits, whilst a third is better known as a winter visitor to the coasts. There may not be any very urgent necessity for protecting these birds at present; but there is no doubt they are disturbed a good deal during the nesting season, and their eggs taken, whilst in winter many individuals are shot in the most wanton manner and left where they fall. This, we regret to say, is a frequent occurrence off the coasts of South Devon, more especially with regard to the Great Northern Diver (*Colymbus glacialis*). We doubt very much if this species ever bred in the British Islands; but the two following Divers do so, and it is respecting these that our remarks are chiefly made. The Black-throated Diver (*Colymbus arcticus*), is by far the rarest and most local species, although we are glad to say it still breeds in considerable numbers, not only in the Hebrides and the Orkneys, but on the mainland from Argyll northwards to Caithness. The Red-throated Diver (*Colymbus septentrionalis*), has much the same range in our islands, frequenting most of the coasts during autumn and winter, and occupying a very similar distributional area in summer, but including the Shetlands. This Diver also breeds sparingly in Ireland, in which country the Black-throated Diver is rarely seen at any season, and has never been known to nest. The Divers probably owe their immunity from persecution to the inaccessibility and remoteness of their breeding haunts; but every year tourists are overrunning the land in ever-increasing numbers, penetrating more out-of-the-way districts, and the Divers should not be overlooked by those most capable of preserving and protecting them. Experience has repeatedly shown us that species once plentiful have very rapidly decreased in numbers, and finally become extinct, when their haunts have been exposed to disturbing influences.

The Great-crested Grebe (*Podicipes cristatus*), is also worthy of mention in the present chapter. We all know that it is a fairly common resident in the British Islands, breeding on the banks of many lakes and meres in England and Wales, as well as in Ireland and the extreme south of Scotland. But we also know that the plumage of this Grebe is held in great request by the furrier, and that the poor bird suffers much persecution in consequence. Once let "grebe" become fashionable for a few seasons—as we hear it is likely to be—and our indigenous stock of birds may soon be greatly reduced, and one of the most handsome bird ornaments of our inland waters well-nigh extirpated. We draw the attention of our bird lovers and bird preservers to the Great-crested Grebe, because we honestly think it requires more protection than it now receives. After all, we cannot be too alert in these matters; for, taught by bitter experience, we know that many another species once common enough is now excessively rare or even lost to our avifauna for ever.

The Grey Lag Goose (*Anser cinereus*) must also be included in this warning list of threatened species. Down to the close of the last century, this Goose—the only species of *Anser* indigenous to the British Islands—bred in more or less abundance in the English Fenlands. Here the wholesale capture of the young birds, together with the drainage and enclosure of its favourite haunts, have been the causes of its extermination. The wonder is that it actually survived in the English lowlands so long. Nowadays it breeds locally and in comparatively small numbers on the Hebrides, and on the Scotch mainland in Ross-shire, Sutherlandshire, and Caithness. Sixteen years ago we had ample evidence of the absolute abundance of this Goose in certain parts of the Outer Hebrides; but now there is a very perceptible falling off, and everywhere the birds appear to be on the decrease. Persecution by man, the robbing of eggs and young, is decimating the indigenous stock, and we seem to be well within sight of their complete extermination. We trust this may be averted, for we should indeed be sorry to see these wild Geese go the way of so many other species. It would be a pity if the semi-domesticated Grey Lag Geese that make their home at Castle Coole in Ireland are to become the sole surviving relics in Britain of a species which possesses so great an interest to naturalists.

We conclude our list with the Goosander (*Mergus merganser*). This remarkable and handsome bird breeds very locally in the Highlands of Scotland, in Sutherlandshire, Argyleshire, and Perthshire. There is even some evidence to suggest that the Goosander is slowly increasing as a British species; but this may, on the other hand, be due to the closer search for its nest. Whatever the facts may be, Scottish naturalists especially should endeavour to preserve this bird from extermination. We have few handsomer native species.

Words of protest might here be written against the cruel and wanton slaughter of many another British bird, at present too common to come within the list of absolutely threatened species. Of these we may mention the Magpie, the Jay, the Hawfinch, the Bullfinch, the Goldfinch, the Sky Lark and Wood Lark, the Nuthatch, the Nightingale, the Woodpeckers, the Kingfisher, the Owls, the Kestrel, the Sparrow-hawk, the Lapwing, and many sea birds. Many, if not all of these birds, leaving all sentiment aside, are absolutely useful to agriculture and horticulture; many of them rank as our most beautiful species. Why need we talk of importing foreign species for their beauty, to adorn our woods and fields, when we have such charmingly arrayed indigenous birds as the Magpie and the Jay, the Goldfinch, the Woodpeckers, the Kingfisher, and the Lapwing?—all of them clad in raiment as fair as that of many exotic species, and all of them endeared to us by the oldest associations. Our Bird Protection Acts, admirable as was the spirit that prompted them, are weak and impotent, because their enforcement is nobody's business. We think the time has come for something stronger than a protest, when about one-fifth of the indigenous avifauna — many of the species of the highest usefulness or entirely harmless—of the British Islands is threatened with more or less speedy extermination! Much has been accomplished already, but more will have to be done; and bird lovers must not, cannot rest until their favourites are in a position of greater security than they are to-day.

Part II

LOST AND VANISHING EXOTIC BIRDS

PLATE VIII.

THE MAMO

LOST EXOTIC BIRDS

THE MAMO

(*DREPANIS PACIFICA*)

WE intend to devote the second part of the present volume to a brief notice not only of some foreign species of birds that have become extinct within the historic period, but of others which are excessively rare, or on the verge of actual extinction, or threatened with extermination if prompt measures be not taken for their preservation. The record of extinction covering the past three or four hundred years is a most lamentable one. Many curious forms have vanished entirely, leaving but the scantiest particulars of their characteristics and habits behind them; respecting others, we have more complete records; whilst some, indeed, have disappeared so recently that their absence yet can scarcely be realised, and of these our information is in most cases more satisfactory.

Our first species carries us away to the fair islands of the Pacific, the home of so many rare and curious birds, doomed, alas! to speedy extirpation. The bird in question is the Mamo, or Pacific Sickle-bill, a species confined to the Sandwich Islands, where it was once very common, but is now so rare that less than half a dozen examples are known to exist in collections. Few as these are, there can be little doubt that they represent the surviving relics of the species, for all recent efforts to find it in a living state have proved fruitless. The extermination of the Mamo cannot be attributed to civilised man. In this case savage man has been the delinquent, destroying the bird for the sake of its beautiful golden-yellow plumage, which was made up into war-cloaks for the Hawaiian kings, and into necklaces for their women. The feathers in request were those from the back of the bird, and to obtain them small bunches were received by the kings as a poll-tax from their poorer subjects, and a regular staff of bird-catchers were employed by the chiefs to augment the supply. Only a few feathers from each bird were suitable, so that many thousands of birds had to be destroyed to furnish the material for a single robe. Formerly, as we gather from Mr. Lucas, the kings, chiefs, and other noble Hawaiians wore these flowing capes or robes whenever they appeared in public on state occasions, either in peace or war, these garments having the same significance and being as eagerly coveted as the ermine and purple in feudal Europe. One of the most gorgeous of these robes was that belonging to Kamehameha I., a powerful king, who not only conquered but united all the islands of the group under his sway. Mr. Scott Wilson, who visited the Sandwich Islands specially to search for the Mamo, says that the manufacture of this great yellow war-cloak had been in progress during the reign of eight preceding kings. "Its length is four feet, and it has a spread of eleven and a half feet at the bottom, the whole having the appearance of a mantle of gold." With the above facts before us, it is not improbable that savage man has exterminated many brilliantly-coloured birds of which we have not any knowledge whatever. There is another allied bird in the Sandwich Islands which has suffered much persecution for the sake of its feathers, the O-o (*Acrulocercus nobilis*), but in this case the procuration of the plumes does not involve death, the coveted feathers (a tuft under the wing) being pulled out, and the bird restored to liberty.

The Mamo was an exquisitely beautiful bird, having most of the upper parts black, with the exception of the lower back, rump, and upper tail coverts, which are yellow; many of the smaller feathers on the wing are yellow, but the quills are black; the tail also is black; the under wing coverts white; the general colour of the under parts dusky, except the vent and the thighs, which are yellow. The total length of this bird was about eight inches; the bill, long, slender, and sickle-shaped, nearly two inches in length.

THE DODO

(*DIDUS INEPTUS*)

ALTHOUGH the precise year in which the island home of the Dodo was discovered is unknown, there can be no doubt that the earliest mention of the bird is contained in an account of the voyage of the Dutch Admiral Van Neck to Mauritius in 1598, published a year or so afterwards. De Bry, the chronicler of this voyage, alludes to the Dodos which were met with on the island, and, so far as we know, seen for the first time by man, as birds "bigger than our Swans, with large heads, half of which is covered with skin like a hood. These birds want wings, in place of which are three or four blackish feathers. The tail consists of a few slender curved feathers of a grey colour." These Dutch pioneers christened the Dodos *Walckvogel*—disgusting or nauseous birds—on account of their poor gastronomic qualities, only the breast being palatable, but probably also because better and more toothsome meat could be obtained on the island—Doves, tortoises, turtles, and fish, which in those days abounded. The poor Dodo, however, was never allowed to remain in peace for long, and the next vessel to reach the island decimated the unfortunate species. This was in 1601, when a ship commanded by Van West Zannen touched at Mauritius, his crew, he tells us, capturing twenty-four Dodos one day and twenty on another, "so large and heavy that they could not eat any two of them for dinner." Van Zannen sailed away with his larder well stocked with salted Dodos: and in ensuing years other ships appeared from time to time to seek supplies of fresh meat; and in less than a hundred years after its discovery the wonderful bird had ceased to exist.

But little definite seems known respecting the habits and economy of the Dodo. That it was a terrestrial species there can be no doubt. François Cauche, who made a lengthy stay upon the island in 1638, furnished more or less trustworthy particulars of the bird, describing its cry as like that of a Gosling, and its single white egg, "the size of a halfpenny roll," laid on a heap of herbs in the forest. It is matter for surprise that so few examples of this curious bird found their way to Europe. Roelandt Savary, a Dutch artist, appears to have made many paintings of the Dodo from life, so that a few captives must have been brought to Holland, and possibly to Austria. About 1638 a captive Dodo appears to have been exhibited in London, Sir Hamon Lestrange recording how he went into the show to see the strange bird that was called by its keeper a "Dodo," and which appears to have been an adept at swallowing pebbles as big as nutmegs. For more than seventy years the Ashmolean collection at Oxford appears to have contained a specimen of the Dodo; but in 1755 it was destroyed, the head and right foot only being preserved, and still in existence in the museum of the Oxford University. A left foot of the Dodo more than two hundred years old is in the British Museum; and a head of about the same antiquity, so far as records go, is in the Museum at Copenhagen.

How long this curious bird had dwelt in peace upon the island of Mauritius, whence it came or whether it had been evolved in the place where man discovered it, are questions concerning the Dodo which will probably never be satisfactorily answered. Its extermination, however, was entirely due to the action of civilised man. When he came upon the scene, he found the Dodo so utterly unsuspicious, tame, defenceless, and even stupid (its name is derived from the Portuguese Dóudo, a simpleton), that its capture was simple and easy enough. From what we know, it must have been a heavy, clumsy bird, quite unable to elude any ordinary pursuit; it was incapable of flight, and could doubtless only waddle in a slow, lumbering manner before its pursuers. The various animals introduced into Mauritius by man also assisted in exterminating the Dodo. When we bear in mind the remoteness of its home, and the comparatively small number of human beings that could visit it, together with the desultory nature of those visits, the extermination of the Dodo was a rapid one; and so quickly and completely did the species vanish, that doubt was widely expressed as to whether the bird had ever existed at all!

The Dodo, with a couple of other allied species which dwelt on neighbouring islands, constitute the family Dididæ, most nearly allied to the Pigeons—a group whose origin may probably date back, according to Dr. Wallace, to early Tertiary times. The Dodo, like the Great Auk, there can be little doubt, owed its flightless condition to the disuse of its wings, probably through long residence upon an island free from enemies, and where aerial locomotion was unnecessary to its existence. As the wings became more and more abortive, the body possibly grew in bulk, owing to a sedentary habit, until at last flight became impossible. We cannot, however, endorse all Dr. Wallace's views respecting the origin of these curious Didine forms; and it seems to us by no means improbable that the group was more widely dispersed at some earlier epoch.

From the numerous paintings of the Dodo which are in existence—some of which we have had the pleasure of examining — we may infer that the bird was a big-bodied one, with short, clumsy legs, enormous head, and huge, ungainly-looking, hooked bill. The body was clothed in loose plumage, the quills alone being rigid, the tail plumose. Its general colour appeared to be dark grey, the breast brown, and the wings and tail white.

THE SOLITAIRE

(*PEZOPHAPS SOLITARIA*)

THE vernacular name now borne by this extinct bird of Rodriguez was originally given to another and doubtless allied species by the French colonists of Bourbon or Réunion. As nothing definite appears to be known of this latter species, the name may be retained by the Rodriguez Island bird, and which was bestowed upon it by Leguat, its earliest historian. The Dodo had been found and exterminated before the present species became known, if we attribute the absolute discovery of the Solitaire to the Huguenot, Leguat; but Professor Newton has shown that earlier explorers may have been familiar with it, or heard of it, if they confused it with the Dodo. Whatever the real facts may be, to Leguat we are indebted for our only knowledge of the characteristics and habits of the Solitaire of Rodriguez.

THE SOLITAIRE

In 1691 he visited the island with the object of founding a colony there, but, fortunately for naturalists, he seems to have devoted more of his time to watching the habits of the Solitaire than to his settlement, which came to an end in a couple of years. The accuracy of his account of this bird had long been doubted, but subsequent researches have confirmed its truth in almost every important particular. From Leguat's interesting account (published in 1708) of the long extinct Solitaire we make the following extract:—

"Of all the birds in the island the most remarkable is that which goes by the name of the Solitaire, because it is very seldom seen in company, though there are abundance of them.

"The feathers of the males are of a brown-grey colour: the feet and beak are like a Turkey's, but a little more crooked. They have scarce any tail, but their hind part covered with feathers is roundish, like the crupper of a horse; they are taller than Turkeys. Their neck is straight, and a little longer in proportion than a Turkey's when it lifts up his head. Its eye is black and lively, and its head without comb or cop. They never fly, their wings are too little to support the weight of their bodies:

they serve only to beat themselves, and flutter when they call one another.

"They will whirl about for twenty or thirty times together on the same side, during the space of four or five minutes. The motion of their wings makes then a noise very like that of a rattle; and one may hear it two hundred paces off. The bone of their wing grows greater toward the extremity, and forms a little round mass under the feathers, as big as a musket ball. That and its beak are the chief defence of this bird. 'Tis very hard to catch it in the woods, but easy in open places, because we run faster than they, and sometimes we approach them without much trouble. From March to September they are extremely fat, and taste admirably well, especially while they are young. Some of the males weigh forty-five pounds.

"Though these birds will sometimes very familiarly come up near enough to one, when we do not run after them, yet they will never grow tame. As soon as they are caught, they shed tears without crying, and refuse all manner of sustenance till they die.

"When these birds build their nests, they choose a clean place, gather together some palm-leaves for that purpose, and heap them up a foot and a half high from the ground, on which they sit. They never lay but one egg, which is much bigger than that of a Goose. The male and female both cover it in their turns; and the young, which is not able to provide for itself in several months, is not hatched till at seven weeks' end. All the while they are sitting upon it they will not suffer any other bird of their species to come within two hundred yards round of the place; but what is very singular is, the males will never drive away the females, only when he perceives one he makes a noise with his wings to call the female, and she drives the unwelcome stranger away, not leaving it till 'tis without her bounds. The female does the same as to the males, and he drives them away. We have observed this several times, and I affirm it to be true.

"The combats between them on this occasion last sometimes pretty long, because the stranger only turns about, and does not fly [flee] directly from the nest. However, the other do not forsake it till they have quite driven it out of their limits. After these birds have raised their young one, and left it to itself, they are always together, which the other birds are not, and though they happen to mingle with other birds of the same species, these two companions never disunite. We have often remarked that some days after the young one leaves the nest, a company of thirty or forty brings another young one to it, and the new-fledged bird, with its father and mother joining with the band, march to some bye place. We frequently followed them, and found that afterwards the old ones went each their way alone, or in couples, and left the two young ones together, which we called a *marriage*.

"This particularity has something in it which looks a little fabulous, nevertheless what I say is sincere truth, and what I have more than once observed with care and pleasure."

Many bones of the Solitaire have been recovered by various investigators, so that the osteology of the species is accurately known, thanks to the unwearying efforts of Professor Newton and his accomplished brother, the late Sir Edward Newton. These bones were mostly procured from caves; but their age seems unknown, although said to belong to a period previous to the colonisation of the island. When we read Leguat's charming and quaint description of this long extinct bird, our wonder increases that so little has been recorded concerning the Great Auk, which dwelt in more accessible parts of the world, and has become extinct so recently, yet, notwithstanding all this, it never met with a biographer that can be compared with the describer of the Solitaire of Rodriguez.

THE PIED DUCK

(CAMPTOLAIMUS LABRADORIUS)

ALTHOUGH the Pied Duck was so well known to American naturalists, and once so common that examples were, according to the testimony of Audubon, Wilson, and other writers, frequently sold in the markets of New York and Baltimore, there are not half as many specimens in scientific cabinets and collections as of the Great Auk. According to Professor Newton, the last example was obtained in Halifax harbour in the autumn of 1852; this specimen was, we believe, until recently in the collection of Canon Tristram, but may now be in the Derby Museum at Liverpool. On the other hand, Mr. Lucas states that no example has been taken since December 1878; but as neither this nor others said to have been obtained between the years 1857 and 1871 appear to be in existence, they cannot well be adduced as evidence. The extinction of the Pied Duck may not have been so sudden as some naturalists suggest. The evidence seems to suggest that the species was by no means a common one in the early days of American colonisation, and that it must have been on the verge of extinction a century or more before that became an accomplished fact. If we admit the possibility of the above surmise, it is easy to understand how the Pied Duck was eventually exterminated, for we know that the remnant of the species was ruthlessly shot down at the breeding-grounds, and the decimation commenced undoubtedly by the Indians during earlier epochs was eventually complete. Again, as Mr. Lucas informs us, "a possible cause for the original depletion may have been the taking of eggs by the Indians, for the Eider, which breeds along the southern coast of Labrador, suffers severely from their depredations. A small dog is trained to hunt through the bushes near the water's edge, the favourite nesting-place of the Eider, while his master silently paddles along close to the shore to note just where a bird is driven from the nest, and in this manner many eggs are taken. Now, if the Labrador [or Pied] Ducks bred over a comparatively small extent of country, near the summer camp of a band of Indians, their original decrease would be readily accounted for." In the case of the present species, we see that not even strong powers of flight were able to save the bird from extinction—a fact which emphasises the importance of including the eggs in any protective measure that may be devised for saving threatened species. It has been suggested that some avine epidemic may have assisted in the work of extermination, but of this there is no absolute evidence. The fate of the Pied Duck may well serve as a warning to us; for it shows that when once a species, or the local indigenous stock of a species in any particular country, becomes abnormally reduced in numbers, its tenure of existence is a weak and slender one, and may be destroyed almost without any assignable cause. Several British species are in this position to-day, in that state in which the Pied Duck was not so many years ago, and their fate may be similar if we do not heed the caution in time. Casualties that under more propitious conditions might only have had a local influence, may now cause complete extinction.

The range of the Pied Duck never seems to have been a very extensive one. In the breeding season the bird appears to have been confined to Southern Labrador, and during winter to migrate along the

THE PIED DUCK

eastern coasts of America as far as Chesapeake Bay. But little has been recorded of the habits of the Pied Duck. Wilson tells us that in his time it was rather scarce on the coasts, and was never met with on fresh-water lakes and rivers. By some gunners it was known as the Sand Shoal Duck, from its habit of resorting to sand-bars. He tells us that its principal food appeared to be shellfish, which it obtained by diving; whilst Audubon was assured that the bird was caught on lines baited with mussels. Wilson writes that nothing more was known of their habits or mode of breeding. This Duck appears to have nested on rocky islands, laying its eggs in a nest similar to that of the Eider.

Whether this species was so closely allied to the Eiders as some naturalists think, seems extremely doubtful. All the male Eiders have more or less green plumage on the head, a characteristic wanting in the male of the Pied Duck. Then the Eiders are birds of remarkably limited migrations, but the subject of the present chapter was noted for its very distinct seasonal movements. Its general style and coloration seem to show closer affinities with the Long-tailed Duck, so that, all things considered, its generic separation seems reasonable.

In the adult male Pied Duck the body feathers and the primaries are black; the rest of the wings, the head and neck white; round the neck a black collar; on the crown a longitudinal black stripe. The female is described as plumbeous grey, slightly darker on the under surface. In size the Pied Duck was said to be about the same as the Long-tailed Duck.

PALLAS'S CORMORANT

(*PHALACROCORAX PERSPICILLATUS*)

ABOUT the same time that the Pied Duck disappeared from the Atlantic coasts of America, Pallas's Cormorant became extinct on the Pacific side of the continent. So far as we know, this, the largest Cormorant of modern times, was an inhabitant of Bering Island, where it was discovered by Steller in 1741, when Bering was wrecked at that spot, the bird being killed for food by the survivors of the fatality. Steller informs us that this Cormorant was very abundant; and as it is evident that the bird was gifted with only moderate powers of locomotion on land as well as in the air, the discovery of its haunt by civilised man was followed by its rapid extermination. In about a hundred years it had become extinct, so far as is known, and all that remains to us are four skins and a small series of bones! It is possible that in this case again extirpation has been largely due to uncivilised man. Dr. Stejneger was told by the natives of Bering Island that the flesh of this Cormorant was exceptionally palatable, and that during the winter, when other meat was scarce, it formed an article of food more highly prized than any of the other Cormorants frequenting the place. It is far from improbable that this noble-looking Cormorant at some distant period occupied the other Aleutian Islands, where it may have been slowly hunted to extinction by the native tribes of those remote regions. From its great size it must have been eagerly sought, for Steller informs us that a single bird—weighing from twelve to fourteen pounds—was sufficient for three of his starving shipwrecked crew. In a somewhat extensive deposit of bones of various mammals and birds on the north-western extremity of Bering Island, Dr. Stejneger found—associated with the bones of Arctic foxes, sea otters, sea lions, and marine birds—a pelvis and other osteological remains of Pallas's Cormorant; whilst on a second visit to the island in 1895, amongst additional bones he obtained another pelvis and a cranium. Of the habits of this Cormorant nothing whatever appears to be known. They were doubtless very similar to those of better-known species of

Cormorants, but the bird does not seem to have used its wings so much, and possibly was most active in the water.

As previously stated, Pallas's Cormorant was the largest of its family known to science. Its general colour was dark green, glossed with blue on the neck and with purple on the scapulars. The shafts of the tail were white. In the nuptial season the neck was adorned with long pale yellow filaments. Round the eyes was a broad ring of bare white skin — hence the specific name.

SOME OTHER EXTINCT FORMS

AS it is impossible within the limits of the present work to deal with each species that has become extinct during comparatively recent times, we may here make passing allusion to a few of the most notable instances, before dealing with a selection of those exotic species that are threatened with more or less speedy extirpation. As we have already shown, island species have suffered most; and many of these are unfortunately surviving forms of avifaunæ that have usually disappeared from other parts of the world, or always been excessively local. Many of these avine species have been lost before any detailed studies of their anatomy and habits have been made, many more are fast going, so that it behoves naturalists and anatomists to lose no time in making themselves acquainted with the various facts. In few other parts of the world has extirpation been more extended and disastrous than in the series of islands in the Indian Ocean known collectively as the Mascarenes. Within this area formerly flourished the Dodo and its kindred; from one or other of these famous islands species after species has disappeared for ever. First we may mention the giant Coot that formerly dwelt in the waters of Mauritius, and the Crested Parrot (*Lophopsittacus mauritianus*), together with that Barn-Owl-like bird, *Aluco sauzieri*; the Dove (*Alectorœnas nitidissima*), as well as the flightless Ralline bird, *Aphanapteryx*. Then from another island in this area has finally disappeared that curious Starling (*Fregilupus varius*), which had its home within the last half-century in Réunion; whilst from lonely Rodriguez—the home, as we have already seen, of the Solitaire—many another has dropped completely out of existence. We have records of what appears to be another form of *Aphanapteryx*, a small Owl (*Athene murivora*), a peculiar Parrot (*Necropsittacus rodericanus*); a Heron (*Ardea megacephala*); and possibly we may now have to add a Paroquet (*Palæornis exsul*). Passing on to the Antipodes, we find extirpation prevailing with grievous frequency, as we have already shown in our opening chapter. Dr. Forbes gives a list of no less than seventeen species which formerly lived on the Chatham Islands, and every one of which has become extinct. Continuing across the Pacific, the same story has to be told; species after species had gone for ever; and here especially the trading collector (and we are afraid the scientific one, too) has played, and is continuing to play, sad havoc amongst these island birds—many of them exquisitely beautiful, and profoundly curious. Onwards again to the West Indies, and still the extermination of birds has progressed, in some of the islands whole groups of certain species having disappeared, and others, according to the most recent information, are quickly following! There is an old proverb that says it is no good crying over spilt milk. Be it so; but the ornithologist may well be excused a tear when he tries to picture what he has lost! Nothing can now recall the many curious and beautiful birds that are gone; but let us profit by the sad experience by endeavouring to save as many as may be of those species still left to us, but threatened with more or less early extermination. A few of these we will now proceed to enumerate.

VANISHING EXOTIC BIRDS

THE CAROLINA PAROQUET

(*CONURUS CAROLINENSIS*)

THIS pretty little North American Parrot is now, alas! a rapidly vanishing species. Bendire, one of the most recent writers on this species, says: "The total extermination of the Carolina Paroquet is only a question of a few more years, and the end of the present century will probably mark their disappearance. Civilisation does not agree with these birds, and as they certainly do some damage to fruit in sections where they still exist, nothing else than complete annihilation can be looked for. Like the Bison and the Passenger Pigeon, their days are numbered."[1]

[1] We hope Bendire and Mr. Lucas have taken rather a gloomy view in this case. Certainly great numbers of these birds have recently

The extermination of this species appears to be directly due to the agency of civilised man. Their numbers, we learn from Bendire's account, from which many of the particulars of this species here given are obtained, have gradually but steadily diminished with the general settlement of those regions frequented by this bird. Audubon, even as early as 1832, tells us that they were not so common as formerly; but even as recently as 1860 they were still comparatively common in the Gulf States, and the Mississippi, Arkansas, and White River valleys. At the present time the Carolina Paroquet is confined to the least accessible portions of South Florida, and very locally to the Indian Territory. As so often happens in cases of this kind, some of the habits of this Paroquet are but imperfectly known, notwithstanding the bird's former wide distribution and abundance. Its favourite haunts appear to be well-timbered valleys and the large cypress swamps so common in the Southern States. It is a very social bird, rarely met with alone, and so fearless that a flock is easily destroyed whilst hovering above a fallen companion, as is the way with certain Terns and other species. Before

been imported into Europe, and hundreds of skins collected in Florida for the Smithsonian Institution (Conf. *Ibis*, 1896, p. 412).

cultivated fruits became so common, the favourite food of this Paroquet consisted of the seeds of the cocklebur and sycamore, as well as those of the cypress and pecan, together with beech nuts, the fruit of the papaw, mulberries, wild grapes, pine cones, and the seeds of the bur grass. Their acquired taste for such cultivated fruits as bananas and oranges, and for Indian corn, has brought down upon them the wrath of the cultivator, and their consequent extermination. Upon the ground this Paroquet is somewhat clumsy, but in the branches it moves about with great agility, climbing here and there amongst the slenderer twigs, often head downwards, reaching and nipping off the buds and berries and fruits on which it subsists, and swinging itself from bough to bough with the help of its strong beak. Its flight is described as undulatory, rapid, and graceful; and so agile are the birds upon the wing, that they dart in and out of the thickest timber with ease even when flying in compact flocks. Their call-note is a shrill *qui* repeated several times, the last utterance being prolonged into a sound like *qui-i-i-i*, and is most frequently heard during flight. When the bird was more plentiful than it is now, it roamed about in flocks numbering hundreds of individuals nowadays it is very exceptional to see more than a score together, and usually small companies of six or a dozen. They are most active in the morning and evening, passing the middle of the day on some favourite tree, hidden amongst the foliage, which assimilates so closely with the colour of their plumage as to render their discovery difficult. Mr. M'Ilhenny says that in fall their food partly consists of the fruit of the honey locust, and that after feeding they retire to drink and to bathe.

There is much difference of opinion amongst naturalists respecting the breeding habits of the Carolina Paroquet. Mr. Brewster (*Auk*, 1889, pp. 336, 337) made many inquiries in Florida concerning its nest, but only three men professed to know anything whatever about it. Two of these—hunters and bird-catchers—described the nest as a flimsy structure built of twigs, and placed on the branches of cypress trees. Confirming these statements, Judge Long assured Mr. Brewster that he had examined many nests built precisely as described above. Formerly he found these Paroquets breeding in large colonies in the cypress swamps. Several of these colonies were composed of at least a thousand birds each. They nested invariably in small cypresses, the favourite position being on a fork near the end of a slender horizontal branch. Every such fork would be occupied; and he has seen as many as forty or fifty nests in a single tree. The nests were similar to those of a Dove, made of cypress twigs, and often so loosely put together that the eggs could be seen through them from below. They ranged from five or six to twenty or thirty feet from the ground. It is difficult to reconcile such testimony with the statements of Wilson and others, who assert that the bird breeds in hollow trees; but we cannot admit that Wilson knew anything about the matter from personal observation, for he tells us that he was unsuccessful in obtaining any information relating to the time or the manner of building of the Carolina Paroquet. He was assured that they bred in trees. Certainly the latter method of breeding is that adopted by most Picarian birds, but possibly this species resembles the Yellow-billed Cuckoo in its methods of nest-building. The eggs of the Carolina Paroquet are said to be four or five in number, and, judging from specimens laid in confinement, to be "white, with the faintest yellowish tinge, ivory-like and quite glossy; the shell is rather thick, close-grained, and deeply pitted."

THE OWL PARROT

(*STRIGOPS HABROPTILUS*)

HERE we have another New Zealand species whose complete extermination seems to be speedily approaching. It is more than sad that a species only known to science some fifty years should be fast vanishing from the ranks of existing forms, and more so when we know that it is one of those primitive forms from which so much may be learnt, and which in this case anatomists do not appear yet to have availed themselves to any exhaustive extent. As is so often the case with weakly and defenceless creatures, the Owl Parrot is said to be chiefly nocturnal in its habits, and probably to this fact may be due its prolonged survival. Soon after its discovery it was said to be an abundant bird in every part of the country, but in nine years from that event it appears to have been exterminated from the settled districts, and is now one of the most local of indigenous species. During the daytime this singular bird is said to secrete itself in crevices of rocks or tree roots, coming out in the evening to search for the plants, seeds, and fruits upon which it subsists. Although it appears sometimes to mount into trees, its usual haunt is the ground. Here it runs about, to some extent assisted by its short wings, which appear unable absolutely to support it in the air. It will thus be seen that the Owl Parrot is a particularly helpless creature in the presence of a predaceous animal. Before civilised man came upon the scene, this helplessness was of little moment, for it could generally manage to elude its natural enemies, the birds of prey. But when man introduces such previously unknown foes as cats, dogs, weasels, and the like, the result must of necessity be a disastrous one for such a terrestrial bird. The Owl Parrot furnishes one more instance of the crass folly of meddling with nature's methods by introducing birds and animals into countries where they are certain to work untold harm, by destroying creatures which might otherwise have survived for ages yet to come.

The Owl Parrot is described by Professor Newton as being "about the size of a Raven, of a green or brownish green colour, thickly freckled and irregularly barred with dark brown, and dashed here and there with longitudinal stripes of light yellow." And again: "Externally the most striking feature of the bird is its head, armed with a powerful beak, that it well knows how to use, and its face clothed with hairs and elongated feathers that sufficiently resemble the physiognomy of an Owl to justify the generic name bestowed upon it."

THE PASSENGER PIGEON

(*ECTOPISTES MIGRATORIUS*)

MORE than ordinary interest attaches to the present species, for it is one that has been captured on several occasions in the British Islands. Of such an interesting and once so abundant bird it is hard to write, that its extermination has progressed so rapidly within the past quarter of a century, that its complete extinction may be looked for during the next decade. Formerly abundantly distributed over the Northern States and Canada, up to near the Arctic Circle, the Passenger Pigeon is now locally dispersed through the deciduous forest areas of Eastern North America — Northern Maine as far west as Northern Minnesota, and Canada up to the shores of Hudson Bay. The rapid decrease of the Passenger Pigeon must be attributed to the direct persecution of civilised man. For more than two hundred years the bird has been sorely persecuted, not only for its depredations on the crops, but for the sake of its flesh. Its numbers, however, were so enormous that even this long-continued decimation appears to have had little effect until comparatively recent years. Wilson estimated its numbers in thousands of millions; whilst Audubon described them in language which competent critics have condemned as exaggerated. No species, however, could withstand the slaughter that has gone on, and the only marvel is that there are any Passenger Pigeons left in America at all! The species, we have reason to believe, can yet be preserved, and it is sincerely to be hoped that American naturalists will see that this is done. The vast hordes that roamed the country even within the past twenty years are gone; their capture is no longer a profitable occupation; and now that the birds are reduced to breeding in scattered pairs instead of in countless flocks, their extermination must certainly be retarded. Bendire informs us that isolated pairs still probably nest in the New England States, Northern New York, Pennsylvania, Michigan, Wisconsin, Minnesota, and a few other localities farther south.

Few birds could have been more gregarious than

the Passenger Pigeon. It was a species that not only migrated in spring and autumn in countless multitudes, but one that nested in colonies of similar abundance. The vast flocks roamed hither and thither in quest of food, and as the season for reproduction approached they selected some woodland retreat, and commenced to nest. Their roosting-places during the autumn and winter were similarly crowded, and the stirring scenes have taxed the resources of many graphic writers to describe them. Particulars of some of the more recent of these "nestings," as they were called, have been given by Bendire. One was in Michigan in 1877 or 1878, near Petosky. This vast breeding colony extended for twenty-eight miles through the forests,—eight miles through hard wood timber, and twenty miles through white pine woods,—every tree of any size throughout that distance containing nests, and many were filled with them. None of the nests were less than fifteen feet from the ground. The birds arrived in this locality to breed in a compact mass five miles long by one mile wide! Compared with this, the breeding colonies of all other known birds sink absolutely into insignificance. One billion Passenger Pigeons were said to have been destroyed at this single "nesting"; this may be an exaggeration, but if a tenth of it represents the actual figures, can we wonder that the poor bird is now becoming rare? Passenger Pigeons are said to be very noisy whilst nesting, the sounds uttered resembling the croaking of frogs, and the combined clamour from a colony can be heard at a distance of four or five miles. The nest of this Pigeon is a slight platform of dead twigs placed on a flat branch or in a crotch near the stem. The two eggs are pure white. It is said that each "nesting" occupies about a month or five weeks. The favourite food of this bird is beech mast and seeds, but since the colonisation of America grain of all kinds is greedily devoured.

The general colour of the upper parts of the adult male Passenger Pigeon is slate grey; the scapulars and some of the wing coverts are brown marked with black; the quills and primary coverts are black, the former margined with white; the central tail feathers are black, the remainder grey marked with white, especially on the outermost. The sides of the neck are violet grey, shot with bronze and green; the remainder of the under parts is vinous chestnut, paler on the centre of the breast and belly, and becoming white on the under tail coverts. The female somewhat resembles the male in colour: but the head is brown, the under parts are greyer, and most of the iridescent hues are absent. She is also slightly smaller. The young are said to be browner than the female, and most of the feathers of the upper parts have pale margins. The throat and centre of the belly are whitish; the neck and breast are brown, with pale margins to the feathers. The total length of the adult male is about sixteen inches, the graduated tail being eight inches in length.

THE CALIFORNIA VULTURE

(*PSEUDOGRYPHUS CALIFORNIANUS*)

THIS magnificent bird, with a spread of wing exceeding even that of the Condor, upwards of ten feet, is another New World species which seems threatened with speedy extermination. As Mr. Lucas writes: "The threatened extermination of the Californian Vulture is indirectly, rather than directly, due to the agency of man, for its suspicious nature has ever rendered this bird difficult to capture, while the breeding-places are in out-of-the-way and often inaccessible localities, and although the Mexican miners of Lower California are said to kill the bird on every possible opportunity, in order that they may use the quills as receptacles for gold dust, the destruction thus caused would naturally be but small. The free use of strychnine in ridding the cattle ranches of wolves and coyotes has caused the disappearance of this bird, which has been poisoned by feeding on the carcasses prepared for the four-footed scavengers." In the face of such facts, it seems difficult to suggest any way in which this noble species can be saved, except by making the practice of strewing poison in this wanton manner illegal. We should like to see the use of poison—and especially strychnine—made illegal in the British Islands; it is a nefarious practice, and its results are far-reaching and disastrous. To their credit be it said, we have known gamekeepers too humane to employ such a method in destroying vermin: although, alas! there are others who without compunction are too ready to place a poisoned egg for the benefit of Jays and Magpies, and any other species that may be tempted to taste the fatal lure. The death agony of these poor unfortunate poisoned birds is too sad to dwell upon.

The California Vulture never seems to have been a very common bird. Formerly this species ranged the country between the Sierra Nevadas[1] and the coast from the Colorado to the Columbia rivers, but the few that are left appear now to be confined to Southern California. Even here the bird is

[1] There is some evidence to suggest that this species is yet fairly common in the Sierra Nevada (Conf. *Ibis*, 1896, p. 412).

said to be extremely rare, and but very few specimens have been obtained of late years. The extreme usefulness of this and kindred species as scavengers should cause them to be carefully preserved. The habits of the California Vulture are very similar to those of other Vultures. The bird is fond of soaring at an immense height, as if surveying the whole surrounding country in quest of food. It is said that this Vulture was formerly in the habit of going up the Columbia River for five hundred miles to eat the dead salmon stranded upon the banks. Its food is almost exclusively composed of carrion, the bird rarely attacking animals unless wounded or weakly. Whether these birds hunt by sight or by smell still seems a moot point. As most readers know, Waterton strongly supported the latter view; possibly the birds find food by the exercise of both senses. Mr. Lucas writes of this species: "Soaring as they do at great heights, these birds command a view over a territory many miles in extent, their keen eyes not only searching the ground below, but keeping a sharp lookout on the behaviour of any of their fellows that chance to be within sight. No sooner does one bird spy a prospective dinner, than another, still farther away, is apprised of the fact by his actions, and in a like manner number two informs a third, so that the good news is rapidly spread, and throughout a vast area the Vultures come hurrying to one point."

This fine Vulture breeds on inaccessible cliffs.

The general colour of the California Vulture is brownish black, with sheeny reflections on the upper parts; the tips to the greater wing coverts form a narrow wing bar; whilst a bar of white extends beneath the wings, and is very conspicuous during flight; the feathers on the upper neck are elongated.

THE HEATH HEN

(*TYMPANUCHUS CUPIDO*)

THIS bird is very closely allied to the now better-known Prairie Hen. The Heath Hen, as it is rather inappropriately termed, for it is an arboreal species, and heath is a plant by no means common in America, formerly had a somewhat extensive distribution, ranging over Eastern Massachusetts, Connecticut, Long Island, New Jersey, and Pennsylvania. Its extermination is probably due to the felling of timber and clearance of land, together with wanton shooting in the nesting season, before game laws came into existence or were so strictly enforced. It is melancholy to know that this interesting species is now limited to an area of about forty square miles on the Island of Martha's Vineyard, Massachusetts. Here, however, this last remnant of a formerly abundant species is strictly protected by law, and it is some satisfaction to be able to quote Brewster's opinion that the bird is in no present danger of extinction. According to recent calculations, there is on an average from three to five Heath Hens to each square mile of ground occupied.

The Heath Hen differs considerably in its habits from its near relative the Prairie Hen. The latter bird is fond of open country, but the present bird is only found in woods. Its favourite haunts are oak scrubs; and here it chiefly subsists on acorns and berries, wandering occasionally to the outlying fields to feed on grain and the leaves of clover, etc. The nest is placed in the oak woods, generally at the base of some large stump, and is a mere hollow lined with leaves and scraps of dry grass. The twelve or thirteen eggs are creamy buff, slightly tinged with green, and unspotted. It is not known that more than one brood is reared in the season.

According to Brewster, the Heath Hen is on an average a pound less in weight than the Prairie Hen, but closely resembles that bird in appearance. The females are practically similar in colour, but the males have the feathers of the neck tufts fewer in number, and the longest ones lanceolate in shape.

THE AMERICAN TURKEY

(*MELEAGRIS AMERICANA*)

THE Turkey of the United States is not the species that was imported into Europe early in the sixteenth century, apparently by the Spaniards, and from which the domestic breed, now such a familiar feature of the English farmyard, has descended. The latter species, possibly the sole progenitor of the tame race, has its home upon the tablelands of Mexico, and, so far as we can ascertain, is still a common bird and likely to remain so. The American Turkey, the subject of the present notice, is unfortunately bordering on extinction; and the same may be said of the sub-specific form found in Florida (*Meleagris americana osceola*). In the early days of American colonisation, the Wild Turkey was common enough and widely distributed throughout all areas suitable to its requirements; but as its ancient haunts became more and more populated

PLATE IX.

THE AMERICAN TURKEY

with white settlers, who not only destroyed the bird, but cleared away its cover, it gradually decreased in numbers, so that the day seems not far distant when a Turkey will be as rare in the United States as it is in an English county! As Bendire remarks, there are plenty of records testifying to the former abundance of the American Turkey throughout the Southern New England States, and of its existence in Southern Maine; but at the present day its total extirpation east of the Mississippi and north of the Ohio River "is only a question of a few years."

The Turkey, like most game birds, is a resident. Bendire writes of its habits as follows: "The Wild Turkey is essentially a woodland bird, and inhabits the damp and often swampy bottom-lands along the borders of the larger streams as well as the drier mountainous districts found within its range, spending the greater part of the day on the ground in search of food, and roosting by night in the tallest trees to be found. From constant persecution, in the more settled portions of its range it has become by far the most cunning, suspicious, and wary of all our game-birds; while in sections of the Indian Territory and Texas, where it has, till recently, been but little molested, it is still by no means a shy bird. These birds feed on beech nuts, acorns (especially those of the white and chinquapin oaks), chestnuts, pecan nuts, black persimmons, tuñas (the fruit of the prickly pear), leguminous seeds of various kinds, all the cultivated grains, different wild berries and grapes, and the tender tops of plants; also grasshoppers, crickets, and other insects. The actions of the gobbler during the mating season, while paying court to the female, are similar to those of the Domestic Turkey, and well enough known to need no description. The call-notes of the Wild Turkey resemble those of the domesticated bird very much; still, they differ somewhat. In feeding, the usual note is *quitt, quitt,* or *pit, pit.* When calling each other, it is *keow, keow, kee, kee, keow, keow,* and a note uttered when alarmed suddenly sounds somewhat like *cut-cut.*"

Of the form found in Florida, Dr. Ralph writes: "Fifteen years ago I found the Wild Turkey abundant in most parts of Florida north of Lake Okeechobee, with perhaps the exception of the Indian River region; but they have gradually decreased in numbers since then, and though still common in places where the country is wild and unsettled, they are rapidly disappearing from those parts in the vicinity of villages and navigable waters." From this gentleman's careful observations we learn that the Wild Turkey has considerably modified its disposition within the past twenty years. Formerly they were somewhat stupid and unsuspicious birds, but now no bird or animal in the country is more alert or more difficult to approach. Although, as we have stated, this Turkey is a resident, it is subject to much wandering about, usually going in flocks of from two or three to twenty individuals. Sadly significant is the fact that, whereas in earlier years large droves might be met with, single birds and small parties have now taken their place.

The Turkey is polygamous, and upon the females devolve all care of the eggs and young. In Florida the pairing season begins as early as February, but in more northern haunts a month or so later. This love season lasts for about three months, and during that interval the males are very pugnacious, seem to lose a good deal of their wariness, and are said to be easily lured by the hunter imitating the call. The hen birds make a scanty nest upon the ground, often at the foot of a tree or beneath the shelter of a bush. This nest is merely a shallow hollow scantily lined with dry grass and withered leaves.

The eggs are from eight to twelve or thirteen in number, ten being an average clutch. Occasionally, however, two hens will lay in the same nest, as many as twenty-six eggs having been found together, one hen sitting upon them, the other standing close by, probably to take her turn in the task of incubation. These are pale orange buff, thickly yet somewhat indistinctly marked with reddish brown.

The familiar Turkey of the farmyard is too well known to require detailed description here.

THE ALDABRAN RAIL

(*DRYOLIMNAS ALDABRANUS*)

ALTHOUGH the present species cannot yet be classed as absolutely rare, its extermination has already commenced, and there is no reason to doubt will proceed rapidly with the spread of the extirpating agents. The present species (closely allied to the *Dryolimnas abbotti*, of Assumption Island) is apparently confined to the small island of Aldabra with the surrounding reefs, and is an admirable illustration of an island form before becoming exposed to those altered conditions of life that have already proved so disastrous elsewhere. In this case the sad work of extirpation is only just commenced, by the cats which have been introduced into the island, and are now running wild to prey upon the ancient avifauna. These cats, it is said, will eventually spread over the entire Aldabran group of islets, and the consequences will of course be most disastrous. Such creatures should never be allowed to enter these small islands at all, peopled as they are with so many interesting types of avine life.

The account of the habits of this Rail, written by Dr. Abbott from personal observation during a residence of three and a half months on the island, are so interesting that we transcribe them in full. This naturalist tells us that the bird is "very common on all the islets of the Aldabra group, abounding on even the smallest, which do not contain more than half an acre, excepting Grand Terre, where it has been exterminated by the cats, which run wild there. Excessively tame and unsuspicious, as well as inquisitive, they run up to inspect any stranger who invades their habitat, occasionally even picking at his toes. Each pair seem to reserve a certain area of jungle for their own use, and chase off all intruders of their own kind. They are very noisy, particularly in the mornings and evenings. The most common note is a clear short cry, or rather whistle, repeated twelve or fifteen times. While whistling, the bird stands erect, with his neck full length and bill elevated, seemingly greatly enjoying his own musical performance. Often a pair joins in a duet, the male and female standing close together facing each other. Another note is a sort of squeak, and appears to be a sign of anger. They also make a series of short grunts, which seem to be a love-note, and is also used in calling up their young. These birds fight among themselves quite fiercely, flying at each other like gamecocks. One frequently gets the other on his back, pinning him down and pecking at him. The battle is quickly decided, and the vanquished gets up and runs away, pursued by the conqueror, who, however, soon halts, and, drawing himself up to his full height, whistles a pæan of victory. They do not seem to inflict much injury upon each other in these combats. Their food is anything organic that they can pick up; they never scratch like fowls, but poke around among the dry leaves with their bills. The few people who lived upon Aldabra told me that the Rails were very destructive in the gardens, and also ate the fowls' eggs; but so far as I myself observed, they do no damage whatever. They are extremely quick in their movements, darting and dodging about the jungle with great activity. They are not absolutely flightless, but use their wings to assist them in leaping, being able to jump and flutter from two to five feet off the ground. In the open they can easily be caught by a man, but once in the jungle no terrier can catch them. On my first arrival in Aldabra, in September, a few pairs were breeding; but the majority did not breed until November and December, when a heavy rainfall occurred. Sometimes the nest is placed in a shallow cavity in the coral rock, being simply a few dry leaves and sticks; sometimes it is a large loose mass as big as a half-bushel basket, a foot or two from the ground, and placed in a dense tangle of grass and euphorbia. In this case the cavity is very deep, only the head being visible as the bird sits upon her eggs. The number of eggs laid, as a rule, is three; one nest contained four; some were said to sometimes contain more, but I did not meet with any. I was unable to ascertain the period of incubation, or to obtain any very young specimens. The hen sits very closely, and can scarcely be driven off her eggs, returning immediately on the departure of the intruder." Bendire describes the eggs of this Rail as follows: "The shell of these eggs is strong, finely granulated, and moderately glossy, and in shape they vary from ovate to elongate ovate. The ground colour is creamy white, sparingly dotted with fine spots of liver brown, vinaceous, and lavender, which are usually heaviest about the larger end of the egg."

We could have no better example of the way in which so many species have been exterminated in various islands after man has appeared upon the scene. Birds absolutely flightless, or only capable of fluttering slowly along close to the ground, tame and unsuspecting as most have been found to be, are utterly helpless in the presence of man, and even more so when their island homes are invaded by such domestic animals as cats and dogs, and such predaceous creatures as mice and rats, that invariably follow man in his wanderings about the world. There can be but one ending, and sooner or later the weakest goes to the wall, and its race dies out completely.

THE KIWIS

(*APTERYGIDÆ*)

THE four species of Apteryx—called "Ki-wis" by the Maoris, and a name by which they are now more familiarly known—must be ranked with some of the most curious and interesting of existing avine forms. They are birds of very local distribution, being confined to New Zealand; and, being flightless, are not only becoming rare, but are doubtless doomed to early extinction. This seems inevitable in islands where the indigenous fauna has suffered so severely since their occupation by civilised man. These wonderful islands seem almost like one of Nature's storehouses, where have been preserved the relics of bygone ages, and where all these beautiful and curious creatures would have been living their harmless lives in peace down to the present day, had man not colonised them, or even introduced so many exotic species with such disastrous results. The Kiwis

PLATE X

KIWIS

are the survivors of a race of birds that has almost entirely vanished from the earth—living examples of an old-time fauna long faded in the mist of ages past and gone. These curious birds vary in size from that of a Bantam up to that of a small Turkey. They appear to have neither wings nor tail, and are clothed with dense hair-like plumage; they have long Snipe-like bills, the nostrils being situated almost at the tip. The nearest surviving relations of the Kiwis are the Struthiones or Ostriches and allied birds, but they differ from these in so many important respects as to warrant their separation into a distinct order. The Kiwis were not known to science until the early part of the present century. Their nocturnal habits will undoubtedly save them longer from extinction, as they are thus far less likely to fall victims to man or rapacious animals. As previously remarked, four species of these singular birds are recognised by naturalists. The species first discovered appears to have been the South Island Kiwi (*Apteryx australis*); the second, from the same island, is the Little Grey Kiwi (*Apteryx oweni*); the third species is the North Island Kiwi (*Apteryx mantelli*); whilst the fourth, the Large Grey Kiwi (*Apteryx haasti*), is found in both islands. By some naturalists this latter bird is thought to be doubtfully distinct; but it is said to be not only larger, and with a stouter bill, but darker in coloration, the bars on the plumage being nearly black.

Sir Walter Buller's account of these birds is certainly the best that has been published, and extracts from this, referring to the North Island species, may aptly be quoted here. He writes, in his classic *History of the Birds of New Zealand*, as follows: "The Kiwi is in some measure compensated for the absence of wings by its swiftness of foot. When running, it makes wide strides, and carries the body in an oblique position, with the neck stretched to its full extent and inclined forwards. In the twilight it moves about cautiously and as noiselessly as a rat, to which, indeed, at this time, it bears some outward resemblance. In a quiescent posture the body generally assumes a perfectly rotund appearance; and it sometimes, but only rarely, supports itself by resting the point of its bill on the ground. It often yawns when disturbed in the daytime, gaping its mandibles in a very grotesque manner. When provoked, it erects the body, and, raising the foot to the breast, strikes downwards with considerable force and rapidity, thus using its sharp and powerful claws as weapons of defence. While hunting for its food, the bird makes a continual sniffing sound through the nostrils, which are placed at the extremity of the upper mandible. Whether it is guided as much by touch as by smell, I cannot safely say; but it appears to me that both senses are used in the action. That the sense of touch is highly developed seems quite certain, because the bird, although it may not be audibly sniffing, will always first touch an object with the point of its bill, whether in the act of feeding or of surveying the ground; and when shut up in a cage or confined in a room, it may be heard, all through the night, tapping softly at the walls. It is interesting to watch the bird, in a state of freedom, foraging for worms, which constitute its principal food; it moves about with a slow action of the body, and the long flexible bill is driven into the soft ground, generally home to the very root, and is either immediately withdrawn with a worm held at the extreme tip of the mandibles, or it is gently moved to and fro by an action of the head and neck, the body of the bird being perfectly steady. It is amusing to observe the extreme care and deliberation with which the bird draws the worm from its hiding-place, coaxing it out as it were by degrees, instead of pulling roughly or breaking it. On getting the worm fairly out of the ground, it throws up its head with a jerk, and swallows it whole."

The food of the Kiwis—it is not known to differ in all the four species—is worms, beetles, and the kernels of berries: pebbles are often found in the stomach of these birds. These birds make little or no nest, laying one or two eggs in a hollow in the ground. These are incubated by the male. The North Island species has been known to lay eggs in captivity, but never successfully to breed. During the breeding season Kiwis are said to be silent. Formerly, when the Kiwis were much commoner than they are now, they roamed about in parties of from six to a dozen, and their shrill cries were a striking feature of the mountainous areas they frequented, sounding near and far in the stillness of the night.

STRUTHIOUS BIRDS:
OSTRICHES, RHEAS, EMUS, AND CASSOWARIES

IT is sad to know that these giant birds, archaic forms with few surviving near relations, and the last remnants of an ancient avifauna, once widely dispersed, are almost certainly doomed to more or less early extinction in a wild state. These great birds, together with the still surviving Kiwis (conf. p. 266) and the long extinct Æpyornithes, of which the fabled "Roc" (of *Arabian Nights* fame) is presumed to be one, form the group of keel-less Aves which are associated under the sub-class Ratitæ. All are flightless, if swift of foot, yet certainly able to hold their own until man's persecution drives them rapidly onwards to complete extirpation. Some of these big birds are continental, and have managed to survive, notwithstanding their flightless state, in areas abounding in carnivorous animals, whilst others under possibly easier conditions have continued to flourish in islands. All probably would have survived for ages yet to come under normal circumstances; but as civilised man has spread over their ancient haunts they are brought into contact with new enemies, which, alas! they are showing themselves powerless to resist. Some of them furnish plumes of great commercial value, and this is incentive enough for the white man, and even his savage representative, to penetrate into their most secluded haunts, and to slay and exterminate without moderation or mercy. Already many areas once occupied by these birds are depopulated, man still continues to penetrate into their less accessible haunts, and sooner or later they will fall from the ranks of existing species.

Of all these birds the Ostrich (*Struthio camelus*), is the most famous and the best known. There is evidence to show that the Ostrich was formerly more widely dispersed than it is now. The probability is that at one time this bird roamed over many of the vast deserts of South-western Asia, although, so far as is known, it is now but a dweller in those of Arabia, occasionally straying into adjoining areas. Its great stronghold at the present time is the deserts and wide treeless plains of Africa, from the Sahara south to the northern borders of Cape Colony. Whether there are three species of Ostrich in Africa or only one is by no means a settled question, nor one which need concern us here, beyond stating that birds from the south have been separated under the name of *australis*, and others from the Somali country in the north-east have been designated by the term *molybdophanes*. The points relied upon seem somewhat trivial ones.

When in a state of freedom the Ostrich is a polygamous bird, sometimes met with in large companies, but more usually in parties of four or five—one male and several females. Canon Tristram states that the Ostriches dwelling in the North African plains and deserts are not so gregarious as those found farther south. These bands of Ostriches do not appear to roam so much as an inexperienced reader might imagine, and under ordinary circumstances confine themselves to a radius of twenty or thirty miles from their headquarters. They are excessively shy and wary birds, never allowing a strange object to approach them very closely, and when alarmed running off at a tremendous pace into the wilderness. At full speed the stride of an Ostrich measures from twenty-two to twenty-eight feet. It is these splendid powers of locomotion that have saved the Ostrich from complete extermination long ago, and stand the bird in good stead at the present time. For speed and endurance the bird may be said to equal almost any other species gifted with powers of flight. As Canon Tristram wrote many years ago in his interesting book on the Great Sahara: "The capture of the Ostrich is the greatest feat of hunting to which the Saharan sportsman aspires, and in richness of beauty it ranks next to the plunder of a caravan. But such prizes are not to be obtained without cost and toil, and it is generally estimated that the capture of an Ostrich must be at the sacrifice of the life of a horse or two. So wary is the bird, and so vast are the plains over which it roams, that no ambuscades or artifices can be employed, and the vulgar resource of dogged perseverance is the only mode of pursuit. The horses undergo a long and painful training—abstinence from water as much as possible, and a diet of dry dates, being considered the best means for strengthening their wind. The hunters of the tribes to the east of the M'zab set forth with small skins of water strapped under their horses' bellies, and a scanty allowance of food for four or five days distributed judiciously about their saddles." During the non-breeding season numbers of both sexes consort together. Another very remarkable fact in the habits of the Ostrich (as well as other Struthious birds) is its association with zebras and antelopes. Mr. Selous records having seen nine Ostriches—four of them males—consorting with an old wildebeest bull. During the breeding season each male Ostrich gathers two, three, or even four females round him, and a place is selected in which the eggs are deposited. It is said that all the hens lay in the same nest, which is a deep hollow in the sand scratched out by the feet of the breeding birds, the excavated material forming a rampart round it. Here thirty or more eggs will be deposited in circles, and upon these the old male broods at nightfall, commencing his task when about a third of the number are laid. The eggs do not seem to be covered during the daytime, the sun furnishing the warmth necessary for their incubation. The hen birds are said to remain in the vicinity of the huge nest to assist in driving off beasts of prey. Outside the nest some twenty or thirty eggs are also laid, and these, some observers assert, are to furnish food for the newly-hatched young. The old Ostriches are extremely careful in visiting the nest not to betray its whereabouts, and will even feign lameness when their helpless brood is threatened by danger. The large eggs are cream yellow. They are good eating, and from their enormous size often form a welcome addition to the traveller's larder—sometimes scanty enough—in these desert solitudes and scrub-covered plains. It is not known that the Ostrich rears more than one brood in the season. The value and use of the plumes of the Ostrich are doubtless known to every reader, and the growing scarcity of wild birds has led to their being kept in captivity and denuded of their feathers at stated intervals. Ostrich-farming is a growing and a profitable industry. This method of obtaining plumes is certainly to be commended, and may prolong the Ostrich's existence as a species; hunters of the wild feathers may eventually not be able to compete remuneratively with the farmers of them.

Far away to the eastwards, in the steamy forests of the Malay Archipelago and Australasia, we enter the home of another type of these gigantic flightless birds. These are the Cassowaries, forming the family Casuariidæ, of which some nine or ten species have been described. They are found in Ceram, New Guinea, New Britain, North Queensland, and elsewhere in the Australian region. Here again there can be little doubt that complete extermination will overtake these curious birds, and perhaps even more speedily than in the case of the Ostriches, Emus, and Rheas, for many of the species are limited in their distribution to islands where colonisation is rapidly spreading. The Ceram Cassowary (*Casuarius galeatus*), is perhaps the best known and the most frequently seen in menageries and zoological gardens. It is confined to the island of Ceram—a small place for such a large species, not quite two hundred miles in length and about fifty miles in breadth in its widest part—where it is said to be still somewhat common. Dr. Wallace thus describes this species: "It is a stout and strong bird, standing five or six feet high, and covered with long coarse black hair-like feathers. The head is ornamented with a large horny casque or helmet, and the bare skin of the neck is conspicuous with bright blue and red colours. The wings are quite absent, and are replaced by a group of horny black spines like blunt porcupine quills. These birds wander about the vast mountainous forests that cover the island of Ceram, feeding chiefly on fallen fruits, and on insects or crustacea. The female lays from three to five large and beautifully shagreened green eggs upon a bed of leaves, the male and female sitting upon them alternately for about a month." It is said, however, that in confinement the cock birds incubate the eggs alone—a custom common to Struthious birds.

Passing on to the mainland of Australia, we find the equally curious Emus, destined, we fear, soon to become totally exterminated. Already the big lonely birds have vanished from all the more settled parts of the country, and as man penetrates still farther afield, the last haunts must in the course of time become depleted, if some means are not devised for their protection. There are two species of Emus known to science, and these are the only members of the family Dromæidæ. The first of these (*Dromæus novæ-hollandiæ*), is apparently confined to South-eastern Australia, having become extinct in the islands that dot Bass Strait and in Tasmania. The second species (*Dromæus irroratus*), is the representative of the Emus in Western Australia. One very remarkable characteristic of the Emus is the curious internal bag or pouch connected with the windpipe. Its use is not yet definitely known. It has been thought to be an organ of sound during the breeding season, whilst some writers have suggested that by filling this pouch with air the bird can better keep its head above water when swimming, for it is well known that the Emus and the Rheas take readily to the water and swim with apparent ease. The Emu, next to the Ostrich, is the largest of surviving birds. Its haunts are open country, expanding plains, and scrub-clothed wastes. It is capable of running with amazing speed, and when brought to bay defends itself by dealing kicks of great rapidity and power. The Emu subsists upon roots of various kinds, herbage, fruits and berries. It is more or less gregarious, and usually met with in small parties. The Emu is probably polygamous. The nest is a big hollow in the ground scratched out by the bird, in which are deposited from nine to a dozen eggs, light or dark bluish green in colour. These eggs are in great request for various ornamental purposes. The cock bird incubates them, the period being ten or eleven weeks. It is to be hoped that our Australian kinsmen will see that the Emu, their national bird, is saved from the extermination which threatens it. This, indeed, should be a comparatively easy task.

The remaining Struthious birds are inhabitants of South America. These are the Rheas, associated in the family Rheidæ by some authorities, constituting a separate Order of others. There are at present three species of Rheas recognised by ornithologists. The earliest to receive a scientific name was *Rhea americana*, a species ranging from Paraguay and South Brazil to Patagonia. The second species to be described, *Rhea darwini*, was named after its discoverer, Darwin, who obtained it during his ever-memorable voyage round the world on the *Beagle*. It is apparently confined to the extreme southern portions of South America, although we should say it is said to occur north of the Rio Negro. It is also asserted that these two species of Rhea sometimes consort together as far north as the Rio Colorado. The third species was named *Rhea macrorhyncha* by Mr. Sclater, and so far as is yet known inhabits the "sertoes" of North-east Brazil. The Rheas are much smaller birds than the Ostrich, but more nearly resemble that bird than the Cassowaries and Emus, although wanting the famous curling plumes. The plumage of the Rhea, unfortunately, has a commercial value so great that it is likely soon to lead to the complete extermination of the bird. Thousands are slain annually, and whole districts have been already depopulated, for the sake of these plumes, which Mr. Harting tells us are known in the feather trade as "vautour." In its general habits the Rhea very closely resembles other Struthious birds. It is more or less gregarious, living in companies on the wide vast pampas, and, like its African relative the Ostrich, frequently consorting—probably for safety's sake—with deer and guanacos. Of its aquatic habits Darwin wrote as follows in his classic record of the *Beagle's* voyage: "It is not generally known that Ostriches [Rheas] take readily to the water. Mr. King informs me that at the Bay of San Blas and at Port Valdes, in Patagonia, he saw these birds swimming several times from island to island. They ran into the water both when driven down to a point, and likewise of their own accord when not frightened; the distance crossed was about two hundred yards. When swimming, very little of their bodies appear above water; their necks are extended a little forward, and their progress is slow. On two occasions I saw some Ostriches swimming across the Santa Cruz River, where its course was about four hundred yards wide, and the stream rapid." The Rhea is polygamous, several females laying about twenty eggs in a large hollow in the ground, and there the cock bird incubates. South America, we now know, during remote ages was roamed by many enormous flightless birds. The Rheas are the only survivors of this distant past, and it is to be hoped that steps will be taken, and that quickly, for their efficient preservation.

SOME THREATENED EXOTIC SPECIES

AS we brought the first part of the present volume to a close by a brief review of a few threatened species of British birds, so may we aptly close the second by a similar notice of a selection of exotic forms which, though still happily surviving in fair numbers, are yet exposed to persecution which may end more quickly and more disastrously than many of us may suspect. There can be no doubt that the vast numbers of skins imported into our islands, as well as into many Continental cities (Paris especially), must prove a very serious drain upon the species represented. What we wrote seven years ago may well be repeated here: "The trade carried on in plumes and bird-skins for hats, muffs, dress trimmings, etc., is enormous. At the present time (we regret the practice still prevails) almost every lady we meet has feathers of some kind on her head-dress or garments; whilst the windows of shops devoted to millinery are quite ornithological studies. The supply of all this feather ornament entails the sacrifice of much bird life; but birds are prolific creatures, and their numbers (in a great many cases) do not appear to diminish in any serious degree at present. Almost every kind of bird is pressed into the industry. Birds of resplendent plumage from equatorial forests—gaudy Parrots, Manakins, Tanagers, Trogons, and Fruit Pigeons— are sent in bales to the markets of the civilised world. Spangled Humming Birds from the New World, like gems of the finest water, come in their millions; Sun Birds from Africa and the East; Ptarmigan from Arctic snows; Snipes and Plovers from northern regions; beautiful Egrets and Herons from southern rivers and marshes—all find a ready sale, according to the ephemeral fashion that may chance to reign supreme." It is comforting to know that in India—whence so many birds came to the plume marts of the West—measures have been taken for the better preservation of many threatened species, mostly common birds in that country, but rapidly becoming rarer from such ceaseless persecution. South America is yet a happy hunting-ground for the bird-hunter, but surely diplomacy need not be very severely taxed to secure a remedy. Nearer home, in Central Europe, much needless slaughter of birds goes on almost unchecked, although here again we think some steps have already been taken, and doubtless better protection will come in time if naturalists will but bestir themselves. North America, again, where so many birds have decreased in numbers, is sadly lacking in protective measures, especially in the Southern States; and here we may suggest a fertile sphere of usefulness for the American Ornithologists' Union.

It is often suggested, not only in popular books on natural history, but in others of more scientific pretensions, that species gradually retire before advancing persecution or colonisation, and the hope is often fondly cherished that threatened species seek remoter and quieter haunts as civilisation advances or as enemies increase. But no greater mistake could be made. The individuals of any species inhabiting certain areas will continue so to do notwithstanding persecution or advancing civilisation, until every one is directly or indirectly exterminated. But we are told this species or that is retiring into less populated localities, finding or seeking retreats remote from man and his works. Nothing of the kind. It is man himself that is advancing over the normal area of the doomed species, extirpating as he goes, and if his colonising movements extend sufficiently far as to include the whole of that normal area, that species is lost. It is one of the most important canons of distribution, that species do not retreat from adverse conditions of life, and one that cannot be too well remembered by all seeking to protect indigenous species from extermination, as well as by collectors and thoughtless sportsmen. It will therefore be seen that the more local a bird may be—in the sense of having a restricted area of distribution—the less capable it is of withstanding prolonged persecution or injurious disturbance. A small area may be soon depleted of its avine treasures, and it is this incontrovertible fact that may well make us pause in the rash persecution of so many localised species, or hasten our endeavours for their safety.

Now, some of the most local of all avine forms are to be found amongst the Humming Birds. These beautiful birds are most abundant in mountainous countries, and many of the species are so extremely local, that a valley, a mountain-top, or an ancient crater is their sole habitat. Many species are

SOME THREATENED EXOTIC SPECIES

confined to various islands. Thus fifteen or more species are found in the West Indies; two species are confined to the Bahamas; Juan Fernandez is the island home of two more; whilst Masafuera and Tres Marias each have their own indigenous species. In this extreme localisation lies the chief danger of extermination. We know that vast numbers of Humming Birds are killed annually for the plume trade, and there is a strong probability that some of these island species, and others dwelling in the most accessible continental areas, may be extirpated. Fortunately, some of the fairest of these feathered gems dwell in remote localities, and where they are scarcely likely to fall victims to the craze for plumes; but others are more readily obtained, and these species seem likely to suffer. We must, however, bear in mind that Humming Birds in most districts they frequent are exceptionally abundant. Most observers agree on this, some saying they are as numerous as bees about flowers. Scores of individuals may often be seen flitting about a single tree. Mr. Henshaw tells us that in a single clump of Scrophularia he counted eighteen Humming Birds "all within reach of an ordinary fishing-rod"; whilst at Apache, in Arizona, he saw two species "literally by hundreds, hovering over the beds of brightly tinted flowers, which in the mountains especially grow in the greatest profusion on the borders of the mountain streams." In the same country Mr. Scott found it no uncommon thing to see from twenty to fifty birds in the air at once. Collectors of Humming Birds for the plume markets, however, do not show any discrimination, and in this way many rare species are thinned out. One of the rarest and most beautiful Humming Birds in existence is the gorgeous *Selasphorus rubromitratus*. Only two examples are known to science, and yet one of these was discovered in a bird-stuffer's shop in San Francisco, mounted for a lady's hat! It is even by no means improbable that species as yet unknown to naturalists find their way into ladies' headgear. If ladies must have Humming Birds, pray let us have them collected with discrimination, and in a way that will not extirpate some of the rarest and most curious and beautiful forms.

Other threatened species are various Herons and Egrets. These birds for the most part breed in colonies, and so wanton and persistent has been their slaughter, not only in Europe and India, but in America, that some districts are almost depopulated. Upon the cruelty involved in this annual massacre we do not care to dwell, and we would fain hope that it has been exaggerated. We cannot understand, for instance, how the old birds are said to be shot down at the nesting-places when their helpless young are already hatched. The delicate plumes of the Egrets are donned for the pairing season, and are consequently at their best before the eggs are actually incubated. As the breeding season progresses, these fragile plumes abrade and are damaged in various ways, so that the plume-hunter is acting against his own interests in shooting the old birds (which we doubt) at a time when the young are abroad and the prized feathers almost worthless.

Of the European species, mention may be first made of the Great White Egret (*Ardea alba*). Although found in more or less abundance throughout Africa, this fine bird has only two important breeding-places in Europe—one of them in the valley of the Danube, the other in South Russia. In the former locality the bird used to be abundant, but the plume-hunters have thinned its numbers most disastrously, and we may fairly class it as a species threatened with extermination in Europe. Its snow-white plumes, adorning the neck and drooping gracefully from the lower back, are the fatal attraction. The Little Egret (*Ardea garzetta*), is likewise much persecuted for the sake of its elegant dorsal plumes. It also is found over most parts of Africa, but it is only a local summer migrant to Europe from Spain in the west to South Russia in the east. It breeds in colonies, especially in the valley of the Danube, and from this district great quantities of plumage have been obtained. This pretty little bird furnishes what is known in the plume trade as "osprey"; nothing to do with the bird of prey of that name, but the delicate rigid filiform feathers that spring in graceful tufts from the back and sides of the Little Egret and some other species. This filmy plumage is the wedding ornament, donned in spring, so that its procuration involves the slaughter of the bird just previous to reproduction. Both these Herons under ordinary circumstances are wary and shy, seldom allowing man to approach them within gunshot; but at their breeding-places much of this vigilance is relaxed, and their slaughter is a comparatively easy undertaking. Some of the American species of Herons have been even more scandalously butchered at their breeding resorts in Florida and elsewhere.

We have already alluded to one species of Vulture that is said to be fast becoming exterminated, and here we may call attention to the partial extinction of a second, the magnificent Bearded Vulture (*Gypaetus barbatus*). This species frequents the mountain ranges of South Europe and Asia, but in many localities is fast becoming rare, whilst in others complete extinction seems to have overtaken it. In some parts of Europe its decrease has been attributed to poison and shooting; whilst in certain Asian haunts the value set upon its plumage has led to its extermination. It is most certainly a threatened, if not a downright vanishing species, and it seems a pity that such a splendid type of raptorial bird cannot be preserved to us.

Some of the species of Petrel are also threatened with extermination. One of these, the Capped Petrel (*Œstrelata hæsitata*), is specially interesting to English ornithologists, because it has been known to visit the British Islands on abnormal flight. Although the distribution of many Petrels is very little known, the present species appears formerly to have resorted to the islands of Guadaloupe and Dominica for the purpose of breeding, but here it seems to have become extinct or nearly so, and our only hope can be that the Capped Petrel has other nesting-places still undiscovered by man. It is said that the disappearance of the bird from Dominica is due to the introduction of a carnivorous animal into the island; whilst the introduced mongoose is also hastening the extermination not only of an allied species, but of a Vulture, the Turkey Buzzard (*Cathartes aura*), which from its terrestrial nesting habits is powerless to save its eggs and young from destruction.

Again, there are many threatened species in New Zealand and the Chatham Islands, to say nothing of the innumerable islands of the Pacific. In some of these remote spots, however, it is consoling to know that the birds are protected to some extent. We believe the French have passed a law for the protection of birds in all islands over which they have authority; in the Chatham Islands, Mr. Chudleigh, we are informed, is doing all he can to preserve the birds, and will not allow them to be shot on his property. It is to be hoped that British influence may also make itself felt, not only in the islands of the Pacific, but on many another shore in remote parts of the world where the birds are being exterminated. A few years ago an important movement was inaugurated for the preservation of the native birds of New Zealand; a memorandum being drawn up by Lord Onslow, the then Governor of the colony, and presented to both Houses of the

General Assembly. In this it was pointed out that many birds were threatened with extermination, from increase of population and the attacks of various predaceous animals lately introduced into the islands. It was suggested that the only efficacious way to preserve these interesting birds from extinction was to set apart certain small islands for their benefit, and to place them under strict protective regulations. What success has attended the endeavour we are unable to say; but it is a step in the right direction, and an example that might be copied with advantage in many other parts of the world.

In conclusion, we may briefly allude to those curious birds, the Penguins (Spheniscidæ)—all of them inhabitants of the Southern Hemisphere, from the tropics southwards to the margin of the Antarctic ice sheet. They breed in colonies, some of these containing many thousands of birds, on the lonely islands of the Southern Seas. These Penguin "rookeries" are, however, becoming much reduced in numbers by the wanton slaughter practised by the crews of vessels sailing on these remote waters. Upon the land Penguins are helpless enough, and may be killed with the greatest ease, being unable to fly. It is simply scandalous that such a curious and interesting type of bird as the Penguin is should be so wantonly and brutally destroyed. But we fear that the weight of British protest is considerably lessened, when we know that a whole community of Penguins was exterminated by the crew of an English man-of-war engaged upon a scientific expedition to Kerguelen Island; boiled down to provide "hare soup" for the officers of Her Majesty's ship *Volage*! It may be urged that Penguins are yet common enough; but we are by no means certain that this is the case respecting some species, and no bird, no species, can survive long such inhuman massacre. It is interesting to remark that in places where birds are judiciously killed for food or feathers, or their eggs systematically collected, they do not appear to suffer to any serious extent. We have only to point to the vast bird colonies of St. Kilda and Iceland, for instance, to confirm these remarks. In St. Kilda the seventy or more people that reside there live upon birds, the egg and bird harvest being gathered every year, with no apparent injurious effect upon the various species congregating there. This has been going on for many years; but the natives are sensible enough to let their birds enjoy a "close time," when they are left in peace to propagate their kind. The same may be said of the vast colonies of Eider Ducks that are protected for their commercial value. These birds are robbed systematically of eggs and down each season, and many adults are killed, yet the Eiders do not decrease, for they are always allowed to rear broods, and the slaughter is by no means indiscriminate. Experience thus teaches us that birds would yield supply enough for all reasonable purposes—either for food or plumage—if judicious care were exercised. It is gratifying to know that Egrets are now being kept in captivity for the sake of their plumes. There is, we believe, an establishment near Tunis where these birds are kept and allowed to breed in a large aviary. The plumes are shorn twice in the year, in May and September, each bird furnishing about seven grammes in the year, valued at thirty-five francs, a sum, after deducting all expenses, which leaves a net gain of some twenty-two francs per bird. This shows how easily we can preserve these beautiful birds from extinction and yet gratify the whim of women for wearing "aigrette." (Conf. *Bulletin Soc. Nat. d'Acclim. de France*, 1896, p. 102.)

www.ingramcontent.com/pod-product-compliance
Lightning Source LLC
Chambersburg PA
CBHW022048230426
43672CB00008B/1110